About the Author

MICHAEL FINKEL has reported from more than fifty countries across six continents, covering topics ranging from the world's last hunter-gatherer tribes to conflicts in Afghanistan and Israel, to the international black market in human organs, to theoretical physics. His work has appeared in *National Geographic, GQ,* the *Atlantic, Esquire, Rolling Stone, Vanity Fair,* and the *New York Times Magazine.* He lives with his family in western Montana.

TRUE STORY

TRUE STORY:

MURDER, MEMOIR, MEA CULPA

MICHAEL FINKEL

HARPER PERENNIAL

NEW YORK • LONDON • TORONTO • SYDNEY • NEW DELHI • AUCKLAND

HARPER ● PERENNIAL

FIRST HARPER PERENNIAL EDITION PUBLISHED 2006.
MOVIE TIE-IN EDITION PUBLISHED 2015.

Designed by Nancy Singer Olaguera

The Library of Congress has catalogued the hardcover edition as follows:

Finkel, Michael.
  True story : murder, memoir, mea culpa / Michael Finkel.—1st ed.
    p. cm.
  ISBN 0-06-058047-X
    1. Finkel, Michael. 2. Journalists—United States—Biography.
3. Journalism—Corrupt practices—Case studies. 4. Longo,
Christian. 5. Murderers—Oregon—Lincoln County. 6. Prisoners—
Oregon—Lincoln County. 7. Child labor—Africa, West—Press coverage. 8. False personation—Case studies. I. Title.

PN4874.F445A3 2005
070.92—dc22
[B]   2004060788

ISBN 978-0-06-058048-3 (pbk.)
ISBN 978-0-06-233927-0 (movie tie-in edition)

15 16 17 18 19   RRD   10 9 8 7 6 5 4 3 2 1

For Jill

What is true lies between you and the idea of you—
a friction, restless, between the fact and the fiction.

—Alastair Reid, "Where Truth Lies"

PART ONE

LIES

ONE

THIS IS A true story. Sometimes—pretty much all the time—I wish that parts of this story weren't true, but the whole thing is. I feel the need to emphasize this truthfulness, right here at the start, for two reasons. The first is that a few of the coincidences in this account may seem beyond the bounds of probability, and I'd like to affirm that everything herein, to the best of my abilities, has been accurately reported: Every quote, every description, every detail was gathered by me either through personal observation, an interview, a letter, a police report, or evidence presented in a court of law. No names have been changed, no identifying specifics altered. Anything I did not feel certain of, I left out.

The second reason is painful for me to admit. The second reason I am making such an overt declaration of honesty is that, relatively recently, I was fired from one of the more prestigious journalism jobs in the world—writer for the *New York Times Magazine*—for passing off as true a story that was, instead, a deceptive blend of fact and fiction.

The firing occurred in February of 2002, soon after I was caught. The following week, on February 21, the *Times* made my dismissal public by publishing a six-paragraph article, on page A-3, under the headline EDITORS' NOTE. The article's final line announced that I would no longer work for the *New York Times*—a line that, I feared, represented the guillotining of my writing career.

Sure enough, within weeks of the appearance of the Editors' Note, I was flogged by the *Washington Post*, the *Chicago Tribune, New York* magazine, an Associated Press report, a dozen different web sites, several European, Mexican, and South American papers, and in a four-minute report on National Public Radio. One writer described my actions as "sleazy," "arrogant," "offensive," and "pernicious," and then concluded that people like me should "burn in Journalism Hell."

I had been informed of the contents of the Editors' Note a few days before its publication, and I'd assumed that responses of this sort might arise. When someone in the fraternity of journalists fails, it's important for the profession to demonstrate that it can be at least as fierce toward its own as it is toward others. So I devised a plan to shield myself. Once the note was made public, I would retreat into a kind of temporary hibernation: I would not answer my phone, or collect my mail, or check my e-mail. The Editors' Note, I figured, would be posted on the *Times*' online edition shortly before midnight on February 20, 2002. I live in Montana, where the local time is two hours behind New York, so I determined that I would commence my hibernation at 10 P.M.

Less than ninety minutes before the cutoff time, my phone rang. I answered. It was a newspaper reporter for the Portland *Oregonian*; his name, he said, was Matt Sabo. He asked to speak with Michael Finkel of the *New York Times*. I took a breath, steeled myself, and said, resignedly, "Well, congratulations. You're the first to call."

"I'm the first?" he said. "I'm surprised."

"Yes," I said. "You're the first. I didn't think anyone would call until tomorrow, after the story runs."

"No," he told me, "the story isn't running until Sunday."

"No," I said, "it's running tomorrow—it's already at the presses."

"But I'm still writing it," he said, "so it won't be in until Sunday."

"What are you talking about?" I said.

"What are you talking about?" he said.

"I'm talking about the Editors' Note," I said. "Isn't that what you're talking about?"

"No," he said. "I'm calling about the murders."

TWO

THERE WERE, it turned out, four murders. The first was discovered on the morning of Wednesday, December 19, 2001, near the town of Waldport, Oregon, in a muddy pond about a mile inland from the Pacific Ocean. It was the body of a young boy, floating facedown a few feet off the rocky shore. A sheriff's lieutenant called to the scene estimated that the boy was between four and six years old. He had dusty blond hair and brownish green eyes. He was wearing only a pair of underpants, white with blue and green pinstripes. He weighed about fifty pounds. He hadn't been dead long, a day or two at most.

There was no identification on the body, and no obvious sign of injury. No one had filed a missing-persons report with the local police. All absentees at local kindergartens and day-care centers were accounted for. No one knew the child's name. A photograph of the dead boy, tastefully retouched—his hair tousled, his eyes shut, his lips slightly parted—was distributed to the local media, in hopes that someone could help identify him.

For a while, the police theorized that a vehicle might have run off the road. A narrow bridge, part of State Highway 34, bisects the pond, which is officially known as Lint Slough, and a city road winds about its perimeter. Maybe the rest of the boy's family, perhaps tourists, were still entombed in a sunken car. This would explain why no one had come forward to identify the body. There were no skid marks on the road, however, and no oil slick in the water, and the bridge's concrete railing was intact.

Even so, three days after the body was found, the local sheriff's office dive team performed an underwater search of the pond, hoping to discover a clue to the boy's identity. Near the cement pylons of the State Highway 34 bridge, in seven feet of water, the divers made a curious find—not a car, but a pillowcase. The pillowcase was printed with characters from the *Rugrats* television cartoon. Inside it was a large rock.

Later in the day, just after noon, the divers made another discovery. This time it was the body of a young girl. She had blond hair and pale blue eyes; she was younger than the boy, but had the same slightly upturned nose and the same rounded cheeks. She, too, was dressed only in a pair of underpants. As with the boy, her body displayed no signs of trauma.

Tied to the girl's right ankle, though, was a pillowcase, this one with a floral print. Inside the pillowcase was another large rock; the weight had held the girl's body under water. The boy, it seemed clear, had been similarly weighted, but had slipped free of his pillowcase and floated to the surface.

The discovery of a second dead child initiated the most extensive criminal investigation in the history of Lincoln County, Oregon. Every child in the two-thousand-person town of Waldport was checked on. No one was missing. Police departments throughout the West Coast were alerted about the unidentified bodies. None could provide a lead. Agents from the Federal Bureau of Investigation searched national databases of missing children. There were no matches.

The mood in Waldport was one of bafflement and fear. Christmas decorations were everywhere, and two children were dead, and nobody knew if a killer was living among them. A few people placed flowers and cards along the railing of the Highway 34 bridge. Once this was started, it seemed the local residents couldn't stop, and soon the bridge was piled with bouquets of roses, handwritten notes, helium balloons, ceramic angels, and a big Barney the Dinosaur inflatable toy.

Some answers were finally provided by a woman named Denise Thompson, who had babysat the children. She had looked after the kids, Thompson told investigators, on Saturday evening, December 15, four days before the first body was found. She'd seen the photograph of the boy, which had been released to the media. Her husband contacted the sheriff's office, and shortly after the girl's body was located, the couple went to the morgue and made the identifications.

The boy, authorities announced, was named Zachery Michael Longo. He was a few weeks shy of his fifth birthday. The girl was his younger sister, Sadie Ann Longo, three and a half years old. Still missing from the family was another sister, two-year-old Madison Jeanne Longo, as well as the children's parents—MaryJane Irene Longo, thirty-four years old, and Christian Michael Longo, twenty-seven. The family lived in the town of Newport, twelve miles north of Lint Slough. The Longos were new to the region; they had moved to Oregon from Ohio three months before.

The whereabouts of the other three members of the Longo family was unknown. No one knew whether they were alive or dead. The babysitter, though, had further information. Denise Thompson told investigators that she had eaten lunch with Christian Longo on the very afternoon that his son's body was found. They'd met that Wednesday at two o'clock—a few hours after Zachery had floated to the surface of Lint Slough—at the Fred Meyer department store, where both Longo and Thompson worked. At the time, Thompson had not yet heard of the boy's discovery, and neither, apparently, had Longo.

In fact, as Thompson informed the sheriff's office, while at this lunch, Longo revealed that his wife had just left him for another man. MaryJane had taken their three children, Longo said, and flown to Michigan. This news came as a shock to Thompson; she and her husband had become friends with the Longos and had not sensed that anything was amiss.

Officers promptly searched the Longos' last known residence, a rental condominium on Newport's Yaquina Bay. It appeared as though the family had abruptly moved out. No notice had been given to the condominium's manager; the rent was left unpaid. The condominium's furnishings were still there, but all of the family's possessions were gone, except for two stuffed animals—a Clifford the Dog and a Scooby-Doo—which were found in a closet. A television set and a microwave oven, both owned by the condominium, were missing. There was no sign of Christian Longo, his wife, or their youngest child.

Many of the Longos' personal belongings, including infant clothing, family photos, women's clothing, and a wallet containing Mary-Jane Longo's driver's license, were found in a nearby dumpster. In the photographs, the Longo children appeared happy and healthy.

The day after the Longos' condominium was searched, divers investigated the waters in front of the unit. It was December 27, eight days after the first body had been found. Just below a wooden ramp leading to docks where dozens of sailboats were moored, the divers retrieved two large, dark green suitcases. One of the suitcases appeared to have a bit of human hair emerging from the zipper. Inside, bent into a fetal position, was the body of MaryJane Longo. She was nude. A mixture of blood and water was seeping from her nose and mouth; later, the medical examiner determined the cause of death to be head trauma and strangulation.

The second suitcase was also opened. Inside was a pile of clothing, a five-pound scuba-diving weight, and the body of two-year-old Madison Longo. There was no blood on her body, and no obvious injury. She was wearing a frog-patterned diaper. She'd been hit on the head and strangled, according to the medical examiner, then placed in the suitcase and dropped into the water.

THREE

THE STORY THAT resulted in my firing from the *New York Times* was supposed to be about child slavery and chocolate. It was assigned by the magazine's editors, who mailed me a package of materials from a London-based humanitarian agency called Anti-Slavery International. In the package was a videotape of a documentary entitled *Slavery*, which had been produced by a pair of highly regarded British filmmakers, Kate Blewett and Brian Woods, and shown on British television.

The film explained that about half of the world's cocoa beans—the primary ingredient in chocolate—are grown on plantations in the central valleys of the Ivory Coast, in West Africa. Many of these plantations, according to the documentary, are worked by teenage and pre-teenage boys who are trafficked in from poorer neighboring countries such as Mali, Benin, and Burkina Faso. Rather than being paid for their work, these boys are enslaved. They labor from dawn to dusk; they are scarcely fed; they are locked each night in cramped, bedless rooms; they receive no medical care and no money; they are frequently whipped.

"When you're beaten," one boy said in the film, according to the subtitles, "your clothes are taken off and your hands tied. You're thrown on the floor, and then beaten—beaten really viciously—twice a day, once in the morning and once in the after-

noon." Runaways who are captured, he added, are sometimes pummeled to death.

The documentary stated that nearly every plantation in the
Ivory Coast uses slave labor. And, said the film, we who live in
wealthy countries and eat chocolate bars are directly responsible.
In one scene, a young boy stared blankly into the camera and, when
asked what he'd like to say to people who eat chocolate, responded,
"They enjoy something I suffered to make. I worked hard for them,
but saw no benefit. They are eating my flesh."

It was a powerful and haunting film, probing what was clearly
an important topic. My editor told me that this was expected to be
a cover article, which meant that the story would receive a consider-
able amount of exposure. I had recently signed an exclusive contract
with the *New York Times Magazine* and had, in the past year, written
three cover stories—one detailing the ill fated voyage of a boat
crowded with Haitian refugees; another about the lives of a group of
Palestinian teenagers in the Gaza Strip; and a third describing the
international black market in human organs.

Before signing on with the *Times,* I'd spent twelve years writing
travel articles and sports stories. My main source of income, for
much of my career, was *Skiing* magazine. The reception I now
received for my *Times* pieces was overwhelming. The CIA invited me
to its headquarters to speak about the situation in Haiti; hundreds
of people, including a congressman, wrote letters in response to the
Gaza story; I was given a $10,000 Livingston Award for being a
"superior" young journalist. I was thirty-two years old, single and
energetic and intoxicated by the attention. I agreed to write the
slave story, and in June of 2001 I flew to Abidjan, the capital of the
Ivory Coast.

Slave practices on the plantations had apparently been ongoing for decades, but the story, as is often the case with news cycles,
had just become a hot one. Packs of journalists had descended
upon the fertile valleys of the Ivory Coast; I met reporters from

France, Germany, the Netherlands, and Mali. The *Chicago Sun-Times* had already run a long story, as had National Public Radio and *Newsweek*. A writer for Knight Ridder Newspapers, the second-largest newspaper chain in the United States, had just spent several weeks in the area.

As regularly happens when a number of journalists are chasing the same story, a well-worn path had formed, complete with guides and drivers and translators. You stepped off the plane, made a phone call to a so-called fixer whose number had been passed to you by a colleague, and everything fell into place. The arrangement was symbiotic. Our work was made easier in a challenging part of the world, and for those on the media path we provided an excellent source of income. In the Ivory Coast the path led directly to the town of Daloa, in the heart of the cocoa-growing region, and once there, to a group representing the child slaves—most of whom had come from Mali—called the Malian Association of Daloa.

The association was staffed by Malians who had emigrated to the Ivory Coast, and its chief mission was to investigate and expose the abuses that befell young Malian-born laborers. I found the group's members extremely helpful. They arranged interviews for me with several teenagers who'd escaped from the cocoa plantations, and these boys told me stories of miserable working conditions, of constant hunger, of brutal beatings.

The young laborers explained how they had been tricked into coming to the Ivory Coast by traffickers promising easy, high-paying jobs. They described how they'd been held captive by the plantation owners and forced to dig holes, plant seeds, and machete weeds under a broiling sun for twelve or more hours a day, six or seven days a week. They related chilling stories about being beaten for no apparent reasons—struck with whips or sticks or bicycle chains. They told me of their harrowing escapes, running through the jungle by night, hiding by day, until they were fortunate enough to find safe harbor with the Malian Association of Daloa.

Officials with the association talked about the prevalence of slavery and called for new child-labor laws in the region and tougher enforcement of existing laws. They told me that the slaves had to either run away from the plantations or work for years before being allowed to leave. Their association, they said, was arranging buses to transport the boys who'd escaped or been released back to their families. They asked for donations to help cover the cost of these bus trips.

It was a tailor-made feature story, an easy winner. A few days after I arrived in Africa, a series of articles with titles like "The Taste of Slavery" began to appear in many Knight Ridder papers, including the *Philadelphia Inquirer,* the *Miami Herald,* and the *San Jose Mercury News*. I read several of them on the internet from my hotel in Abidjan.

The Knight Ridder stories, while not as dramatic as those presented in the British documentary, reported that there were perhaps tens of thousands of child slaves toiling on cocoa plantations in the Ivory Coast, and that most chocolate bars were therefore tainted with slave labor. According to the articles, some of the slaves were as young as nine years old, and many were routinely and savagely beaten. A boy named Aly Diabate was featured in one of the stories. Diabate said that he was not yet twelve years old when he was tricked into working on a plantation. He said that he had labored for a year and a half and was whipped nearly every day.

The stories were superbly written, peppered with illuminating details and heart-wrenching quotes. They had such a profound impact on a U.S. congressman named Eliot Engel, a Democrat from New York, that he read much of the Aly Diabate story on the House floor. ("Aly was barely four feet tall when he was sold into slavery, and he had a hard time carrying the heavy bags of cocoa beans. 'Some of the bags were taller than me,' he said. 'It took two people to put the bag on my head. And when you didn't hurry, you were beaten. . . . The beatings were a part of my life.'") Engel eventually sponsored an amendment to a bill that included $250,000 to develop standards for labeling chocolate as slave-free. It was passed easily by the House. The authors of the Knight Ridder series later won the $10,000 Livingston Award—the same prize I had won the previous year—and also the Polk Award for international reporting, a journalism honor probably second in prestige only to a Pulitzer.

My plan was to write a piece very much like the Knight Ridder ones. But after about a week in the Ivory Coast, I began to sense that the story was not quite what it seemed. There was something unsettling, I felt, about a few of the members of the Malian Association of Daloa. For one thing, I didn't like the way that some association officials aggressively solicited money. One vice president, Cisse Samba, proudly showed me a two-inch-thick stack of journalists' business cards and insisted that every one had "contributed" to the association.

In journalism, there's a hard-and-fast rule about paying people who are quoted or who provide key information: You can't do it. But the ethics governing the treatment of ancillary helpers such as interview facilitators, cultural liaisons, or city guides are not at all clear. In order to uncover a good story, I'd learned over the years that it was often necessary to present a timely gift, or grease a few palms, or pick up a hefty bar tab. In Haiti, I'd once paid for the rental of an electric generator, a professional DJ, and several cases of beer so that I could entertain some people whose help I needed in researching my article. In Gaza, I made sure to buy all my provisions from a certain shop because I wanted permission to interview the owner's teenage son.

Soon after I arrived in Daloa, a Malian Association vice president named Diarra Drissa spent several hours introducing me to various interview subjects, then listening in and occasionally aiding my translator as I conducted interviews. When I said good-bye to Drissa at the end of the day, he refused to shake my hand, instead telling me, "Our business is not done." I was uncomfortable with the situation because I considered Drissa himself a source for my story, but I wanted more introductions, so I gave him a generous tip for that part of the world—$30 in cash, as well as my raincoat, which he'd been eyeing all day.

Later that evening Drissa showed up at my hotel, furious, saying that I had not paid him nearly enough. Other journalists, he insisted, were far more generous. I wasn't sure whether he was telling the truth, and when I refused to give him further payment, he announced that he would never work with me again, and marched off.

Lying in bed that night, I thought about the incident and replayed, in my mind, the interviews Drissa had organized. Something was off. Even with allowances for language barriers—most of the laborers spoke only Bambara, a main tribal language in Mali—many of the stories the child slaves told me sounded remarkably similar. A level of detail seemed missing. The narrations felt overly rote and unemotional for such disturbing experiences. No matter how I'd phrased the questions, the answers I heard had a faint whiff of falseness about them.

The next day, I worked with a different Malian Association official. He arranged for me to speak with a former cocoa-plantation worker named Adama Malé. The interview took place in the association's cramped cinder-block office. At one point, Malé began describing a failed escape attempt that had occurred several months before he was finally released by the plantation's owner.

This is what Malé said to me, as related by my translator: "I tried to escape and I was caught and beaten. When they catch you they take your clothes off and tie your hands." At this point, Malé stood up from the wooden bench he'd been sitting on and demonstrated, pressing his wrists together in front of him and leaning slightly forward, as though looking for something on the ground. I'd seen the same reenactment from several other boys. "I was hit with a fan belt from a motor. On my back." I asked him if he bled and he said, "Yes, there was much blood."

As he said this, I thought about the mosquito bite on my right hand. I'd been bitten days earlier, in the fleshy spot between my thumb and forefinger, and had picked at it until it bled. Though I'd rubbed antibiotic lotion on the spot, it had become infected and was now a yellowish purple mass, inflamed with pus. Such things happen in a tropical environment.

If my insignificant bite, carefully tended, appeared so bad, I could only imagine what a person would look like if his back had been ripped open by a fan belt and he had no access to medical supplies. In the *Slavery* documentary, one child described what became of such lesions. "After you were beaten," he said, "your body had cuts and wounds everywhere. Then the flies would infect the wounds, so they'd fill with pus. You had to recover while you worked."

I asked Malé, politely, if he would mind taking off his shirt. He was wearing a threadbare oxford that had likely been donated by an aid agency. He looked at me with partly hooded eyes—we were both embarrassed—and then, still standing and facing me, he began to unbutton his shirt. He was tall and painfully thin, with beautiful, delicate fingers. When he reached the last button, he pulled his arms through the sleeves and held the shirt balled up in his hands.

I asked him to please turn around. My translator translated, and Malé turned slowly around. I really wasn't surprised. His back was as smooth as marble. There was not a nick, not a scratch, not so much as the slightest shadow of a scar.

FOUR

THE LONGO FAMILY murders, according to investigators, all probably occurred shortly before dawn on Monday, December 17, 2001. There was one possible witness. A man named Dick Hoch had seen someone at the spot where the two older children were dumped. That Monday morning, at about 1:30 A M , Hoch had been heading to the coast, on his way to work—he removes beach sand that has blown onto people's driveways—when he saw a reddish minivan stopped on the State Highway 34 bridge just outside the town of Waldport, Oregon. The van was facing east, away from the Pacific Ocean.

Hoch, who contacted the sheriff's office after he learned of the dead children, said he was concerned that the vehicle was disabled, so he pulled his pickup truck over to assist. There appeared to be a lone white male in the van, Hoch said, though he could not see clearly because the van's interior lights and headlights were both off. It was a cold morning, a few degrees above freezing, the streets glazed with rain. Hoch asked the man if he needed help, and the man said that he did not—his engine light had flashed on, he explained, and he was just checking it. Hoch drove off, and he watched through his rearview mirror as the van headed across the bridge and down the road.

Christian Longo's vehicle happened to be a maroon Pontiac

Montana van. That Monday afternoon, Longo drove the van to the Fred Meyer department store and worked, according to the store's records, from 2 P.M. until 11 at night. Tuesday was his day off, and by midmorning he had driven about a hundred miles north, to the Portland outskirts.

In the months before the murders, the Longo family had lived a rather chaotic existence—they'd moved from a rental house to a hotel room to another hotel room to a condo in just the past few weeks. Before that, for a fortnight, they'd lived in a tent, and before that, in an old warehouse. The van was one of the family's few constants, and its interior was a jumbled collage of their lives. Later, when officers enumerated every item in the vehicle on search-warrant forms, the list required twelve pages. There was a scooter, a miniature car, a stuffed animal, a sippy cup. There were videotapes—*Toy Story 2*; *Time for Counting*; *Cartoon Crack-Ups*—and, installed over the van's rear seats, a pull-down video monitor. There was camping gear, sunscreen, diet pills, lipstick. There was a children's book, *Zoo Book*, and an adult book, a Lisa Scottoline legal thriller entitled, curiously, *Running from the Law*.

Longo drove his van to the Town & Country Dodge dealership, in the Portland suburb of Wilsonville. He parked in the dealership's outdoor lot and pulled the license plate out of the metal frame at the rear of his van. The plate, from Michigan, where Longo and his family had lived for several years, read KIDVAN. Longo also grabbed his tool kit, his cell phone, and a folder of personal documents, then entered one of Town & Country's large indoor showrooms.

Some of the vehicles in the showroom had keys inside them. Many also had legal plates. A salesman approached, but Longo waved him away, saying that he didn't need any help. The salesman wandered into another showroom. Earlier in his life, Longo had owned a green Dodge Durango, and here in the Town & Country showroom was another green Durango, nearly new. Longo

climbed inside. The keys were already in the ignition. It was unbe-
lievably simple. He started the car and drove it over the weight-
sensitive trigger on the showroom floor, which activated the
automatic garage door. The door opened, and Longo drove out.
Nobody saw him.

When employees of Town & Country noticed that the Durango
was missing, they figured a customer was merely taking it for a test
drive. Not until the following morning—Wednesday, December 19,
the same day Zachery Longo's body was pulled from Lint Slough—
did they contact the police.

After stealing the car, Longo drove back down to the waterfront
condominium he was renting in Newport. That evening he went to
a Christmas party. The gathering was held at an Italian restaurant
across Highway 101 from the Fred Meyer department store. It was
hosted by the staff of Fred Meyer's in-store Starbucks, where Longo
had worked until his promotion to the home-furnishings section
three weeks before. For the party's gift exchange, Longo brought
along an unopened bottle of his wife's perfume.

The next day, Wednesday, Longo arrived at work on time for
the 5 A.M. shift. He informed the home-furnishings manager, Scott
Tyler, that his wife and kids had moved away, and that he could
now work whatever hours Tyler needed. This was also the day he
had lunch with Denise Thompson and told her that MaryJane and
the children had gone to Michigan.

Longo worked the same shift, 5 A.M. to 2 P.M., on Thursday. On
Friday, he was scheduled for a late shift that started at three in the
afternoon. Longo was usually punctual, so when three o'clock
passed and he hadn't come in, his manager tried to call him at
home. Nobody answered, and Longo did not show up at all.

The desire to escape Newport, Longo later recounted, had
come upon him after work on Thursday, while he was at the gym.
He'd just arrived there when the radio station on the gym's sound
system announced that a young boy had been found dead in the

water. The station gave the boy's description and said that he had not been identified. As soon as Longo heard this, he felt nauseous and hurried to the bathroom. He splashed water on his face until he settled down. Then he played volleyball for two hours with a few friends from work.

He didn't sleep much that night—he stayed in the condominium and drank a couple of beers and eventually dozed on the couch. Early the next morning, Longo packed the stolen Durango with most of his belongings, as well as the television and microwave that came with the condo, and drove out of town. He wasn't sure where he was going, he later explained, except that it had to be where no one would recognize him. He drove east, toward Interstate 5, and about fifty miles from Newport he realized it was a Friday—payday at the Fred Meyer. He turned around, drove back to Newport, picked up his $230 check, cashed it at the store, and left again.

Longo returned to the interstate. He didn't know whether to go north, toward Seattle, or south, toward San Francisco. He took the first on-ramp he came to. It was south. The farther he got from Newport, he later said, the better he felt, so he kept driving, past the Cascade Mountains and the Klamath National Forest and the Sacramento Valley. He drove six hundred miles, then exited the highway in Sacramento. He parked in a residential area and slept in the Durango.

Longo reached San Francisco around noon on Saturday, just about the time that divers found his daughter Sadie at the bottom of Lint Slough. He stopped by a bookstore and bought a guidebook to inexpensive San Francisco hotels and another on local campgrounds. He decided to take a room for two nights, at $22 a night, at the Fort Mason Youth Hostel, adjacent to the Marina District. He hadn't eaten a full meal in two days, so he walked to a Safeway and purchased bagels, ramen noodles, cheddar cheese, and Triscuits, then ate them in the hostel's kitchen.

He'd spent nearly half his paycheck on gasoline during the drive down, and he had nothing in the bank. In a matter of days, he'd be out of money. The next morning—Sunday, December 23— he filled out a job application at the Starbucks on Union Street. By this time, Denise Thompson had spoken with sheriff's officers in Newport and had identified the bodies, and Longo was a murder suspect, pursued by federal authorities.

The Starbucks application he completed in San Francisco was later recovered by the FBI. On it, Longo wrote his name, accurately, as Chris M. Longo and said that his social security number was 315-02-4297, which is correct except for the last digit. He listed as a reference his manager at the Fred Meyer Starbucks. For callbacks, he left his cell-phone number. The manager of the Union Street Starbucks said he'd likely have a job for Longo in a few days.

Back at the hostel, Longo checked the news online. He pulled up the web site of the *Oregonian* and saw a headline about the two bodies found in Lint Slough. Longo clicked on the story, and up came a retouched photo of his son.

Longo fled the hostel, climbed into the Durango, and drove away. He parked the car and, he later said, cried as hard as he'd ever cried in his life. He decided that he couldn't return to the hostel—he didn't want to be around people. Instead, he drove to a beach by the Presidio, in the shadow of the Golden Gate Bridge. He sat in his car, gathering his nerve. His intention was to walk to the center of the bridge and jump off. He sat in the car for hours, envisioning stepping onto the bridge. But he couldn't do it. He never even got out of the car.

That night, he parked the Durango on a San Francisco side street. He hung towels and shirts over the windows, crawled into the backseat, covered himself with his leather jacket, and tried to sleep. In the morning, he used the bathroom at Golden Gate Park, then drove to the San Francisco Zoo and sat for most of the day in a secluded spot in the Africa section. That evening, Christmas Eve, he parked for the night on a steeply sloped street. He could hear, he later said, the sounds of a Christmas party emanating from an apartment above him—people playing the piano, people singing carols.

On Christmas Day, a Tuesday, Longo left the Durango and began walking. Nearly every place was closed, except a movie theater. He bought a ticket and watched *Ali*. When it was over, he didn't want to leave, so he stayed and watched it again. Then he walked some more. A Walgreens drugstore was open, so he went in and wandered aimlessly through the aisles. Then he walked again. At a Chinese restaurant, he ordered a noodle dish to go and walked back to his car and ate it there.

He drove around the city on Wednesday and eventually found himself at a park called Lands End. He followed a trail for a few miles until he reached a set of cliffs overlooking the Pacific Ocean. He sat with his feet dangling from the edge and again wanted to end his life. He stood up, backed away a few feet, and then ran to the lip of the cliff, but he couldn't fling himself off.

The next day—Thursday, December 27—FBI agents got the break they were hoping for. That morning, the manager of the Union Street Starbucks decided to check the reference on Longo's employment application. The manager called the Newport Fred Meyer, and an employee there, upon hearing Longo's name, contacted the police. The police notified the FBI. The FBI, with the assistance of Starbucks officials, swiftly devised a sting operation.

In the meantime, Longo had determined that he needed to leave the United States. Wednesday afternoon, the day before the FBI learned of his Starbucks application, he drove to a Kinko's and used their internet service to book a flight to Cancún, Mexico, leaving late that night and returning a month later. He made the reservation under his own name and paid using a credit-card number from a receipt he'd pocketed several weeks earlier while working the cash register at Fred Meyer.

The FBI's plan was to apprehend Longo at the Union Street Starbucks. A Starbucks manager left a message on Longo's voice mail, requesting that he come in for a job interview on Friday, December 28. Though Longo didn't return the call, that morning, several FBI agents were sprinkled anonymously among the usual crowd. The interview time came, then passed. There was no sign of Longo.

The FBI was too late: Longo had already left the country. After booking the flight, he'd driven to the San Francisco airport. On the way there, he pawned the microwave and TV he'd stolen from the condo in Newport, for which he received $90. At the airport he checked in, without incident, for American Airlines Flight 1048, San Francisco to Dallas. He waited in Dallas, then transferred to the early-morning nonstop to Cancún. He'd traveled to Mexico four times before, all of them with MaryJane and twice with his children. Usually, Longo later said, he was a talkative passenger. This time, he didn't speak with anyone on either flight.

When Longo failed to show up for his Starbucks interview, the FBI switched tactics. They decided to make the hunt for Longo both a nationwide affair and a public one. Charles Mathews, the chief FBI agent in Oregon, appeared on NBC's *Today* show and on CNN's *Live Today* to explain the charges against Longo and ask for any information the public could provide. He said that Longo might be driving a green Dodge Durango with a KIDVAN license plate.

Hundreds of tips were phoned in, some from as far afield as Florida and Iowa, with many callers saying that they'd spotted the plate. Nothing, however, was helpful—KIDVAN plates had been registered in at least twenty-five states. Longo's parents, Joe and Joy Longo, who live in Indiana, issued a statement to the press, pleading for Christian to turn himself in. Longo never heard his parents' appeal.

On January 6, nine days after the futile Starbucks sting, the

Dodge Durango was found. Two San Francisco police officers spotted it in a short-term parking garage at the airport. Inside the vehicle was a laptop computer, a cell phone, a box of Triscuit crackers, some cheddar cheese, two empty bottles of wine, and a KIDVAN license plate, which he'd never attached to the car.

Over the next few days, Longo was placed on the FBI's Ten Most Wanted Fugitives list and profiled on the television show *America's Most Wanted*. Longo, who had never previously been accused of a violent crime, was now on the same list as Osama bin Laden. His wanted poster called him "armed and extremely dangerous" and also mentioned that he "has been known to frequent coffee houses." John Walsh, the host of *America's Most Wanted*, said this of Longo: "He's very, very charming. He's very, very smart. He's very calculating. He's really, really good at disappearing."

A $50,000 reward was offered by the FBI for information leading to Longo's arrest, but the bureau also announced that Longo had evidently caught a flight to Mexico, to the resort area of Cancún. Spanish-language wanted posters, said the FBI, were currently being circulated across eastern Mexico, but in truth, the agency admitted, nobody really knew where Longo was.

FIVE

AS ADAMA MALÉ stood in the tiny room in the Ivory Coast town of Daloa, his scarless back turned in my direction, a shock of understanding came over me. I am not a natural skeptic. I tend to believe what people tell me, especially if it confirms my expectations. But there was clearly something wrong here. At that moment, I changed my tack—rather than searching for slaves, I was now looking for liars. And then, as if a code had been cracked, everything suddenly made sense.

"Where are your scars?" I asked Malé.

The young man twisted around and faced me. His eyes were still half-lidded and shy. "The scars have disappeared," he said. He put his shirt back on.

There were a half-dozen other boys hanging around the Malian Association's compound, sitting in the shade, swatting at flies. Most of these boys had told me that they, too, had been beaten. Some had escaped from plantations only days before. I walked over to them. "Can any of you," I asked, "show me scars from being beaten?"

No one said they could. "It doesn't need to be a big scar," I said. "Just a little one."

The boys shook their heads. "All the marks have healed," one explained to me.

Then Adama Malé perked up. "My friend was beaten," he said. "He has marks."

"Where is he?" I asked.

"He's still on the plantation."

"Okay," I said. "Let's visit him."

"When?" said Malé.

"Right now," I said. I had a rental car and driver at my disposal.

"No," said Malé. "We can't. He has left the plantation and gone home."

Later that day, I had a discussion with a fellow journalist, a Paris-based filmmaker who was gathering footage for a French version of the British documentary. His name was Nils Tavernier. He told me that he'd filmed dozens of interviews and had listened to many anguished tales. I asked Tavernier if the stories had begun to sound repetitive, and he admitted that, in some ways, they had. I told him about Adama Malé, and how his back had no scars. I asked him if he'd filmed anyone who had shown him physical confirmation of being whipped.

The question seemed troubling to Tavernier. He was quiet for a while, perhaps attempting to recall all the scenes he'd shot. I'd been in Africa only a few weeks; he'd been back and forth from France for the past year. His answer was unambiguous. "I have never seen evidence of one person being beaten," he said.

In the documentary *Slavery*, there *is* one boy who has horrible scars across his neck and torso and arms. According to the video, several months had passed since he'd been rescued from a cocoa plantation. This one boy's scars are pictured, repeatedly, in the film—often accompanied by a sound track of a cracking whip—but no one else's wounds are shown.

I thought a great deal about this one beaten child. His scars seemed to prove that the injuries to a boy who had been whipped, even months earlier, were horrendous and unmistakable. They did not simply disappear. This strongly implied that the boys from the

Malian Association of Daloa were lying to me. Perhaps, I thought, they were being coached to tell such stories. It was possible that this was being done so that journalists would have powerful material. This would entice more journalists to visit. Everyone would be asked for donations, and the stories would generate further contributions. It was an efficient way for the Malian Association to raise money. If this scenario were true, I realized, then the type of abuse the British documentary says is commonplace might in fact be quite rare. It might hardly exist.

I needed to investigate the plantations myself. As a writer for the *Times*, I was fortunate enough to have a generous expense account and several weeks' time to secure my story. Most journalists were not so lucky. They worked on tight budgets and tighter deadlines. The cocoa plantations of the Ivory Coast are located in the midst of dense jungles that are difficult and expensive to access. It made more sense, time-wise and money-wise, to allow the Malian Association of Daloa to handle logistics.

But I had rented a four-wheel-drive, high-clearance vehicle and had hired an experienced driver. My translator was an expert; his English was perfect. So we drove beyond the broken pavement of Daloa's city streets and onto the packed-dirt secondary roads, past clusters of mud huts and banana trees and young men who'd killed bushrats with their slingshots and were holding the rodents out, hoping we'd want to buy some meat.

Then we entered the jungle. Here, the route was little more than a wide footpath; the only previous tire tracks had been left by bicycles. The grass grew taller than the car and arched over the path on both sides, nearly forming a natural tunnel. We drove for more than an hour, flattening the grass beneath our wheels. The soil seemed to have been dyed bright red, like a bolt of fabric; termite mounds rose sharply skyward.

Cocoa plantations in the Ivory Coast are mostly small and independently owned. There are an estimated six hundred thousand of them, many of which are extremely remote little islands in a vast green sea. We drove the paths until they became too crude to drive upon. Then, to reach the plantations, we started to walk. The plantations are often a mile or two apart, with wilderness in between. My translator and I walked for hours, conducting interviews at each stop. We drove back to Daloa in the evening and returned, via a different route, the next morning. We went back and forth for the better part of a week.

In the British documentary, the president of the Malian Association of Daloa, a middle-aged man named Diabe Dembele, said this: "You'll find slavery on at least ninety percent of the plantations." During my walks, I visited more than twenty-five plantations. I tried to arrive unannounced, so that no one would have time to hide anything. I spoke with more than sixty workers who'd been brought to the plantations from neighboring countries. One of them admitted that he was fourteen years old, another said he was sixteen, and all the other workers, except for the children of plantation owners, told me they were at least eighteen.

Not one of the sixty or more workers I spoke with on my walks said that he had been beaten. None appeared ailing or badly injured, though one worker had cut his own foot while swinging a machete and was wearing a dirty bandanna wrapped around the wound. A few workers admitted they were homesick or wished the food tasted better or the labor were not so difficult, but none mentioned that they felt afraid or were planning to run away. Some said they'd heard rumors of beatings, but no one told me that they'd actually seen a worker of any age being whipped. An article printed in the *Chicago Sun-Times* several weeks before I arrived quoted a Malian diplomat: "It was rare," he claimed, "to meet a child who had not been beaten."

This isn't to say that the living conditions on the plantations could even remotely be described as adequate. The workers slept several to a room on the bare floors of leaky mud-brick buildings. When

I first saw a plantation's living quarters, I thought the structures were chicken coops. None had electricity. The food, mostly made from cornmeal, could in no way fulfill the laborers' nutritional requirements. Their clothing was tattered, their footwear insufficient. They had no chance to leave the plantations and visit the city. They had no opportunity for schooling.

On the majority of the plantations I visited, though, the food and the living conditions of the owner's family were similar to those of the laborers. The plantation owner's children usually didn't go to school, either. The owner and his wife and their children also slept on the floor, also ate cornmeal, also worked from sunup to dark. Men and boys in the owner's family toiled in the fields; women and girls chopped firewood and pounded corn and cooked meals. Life was short and hard for everyone.

Several plantation owners spoke with me at length. They told me about people called "locateurs"—men who bring farmhands from poor villages in Mali and other countries to plantations in the Ivory Coast. Yes, the plantation owners told me, they did pay the locateurs for the workers, and yes, this purchase price was taken out of each worker's salary. They were very open about these transactions; two owners even showed me their accounting books.

Most workers, I learned, were paid a monthly salary of ten thousand Central African francs, the equivalent of about fourteen dollars. To pay back their purchase price usually required three months' labor. So the first year's salary was $14 times nine months: $126. This is more money than most people in Mali earn. The workers were paid once a year, in the fall, when the cocoa beans were sold—the only time most owners were paid. If a laborer quit or ran off before completing a full year, he was not paid at all. At the end of the year the worker could take his money and leave, or elect to stay for another annual cycle.

It was true, the owners admitted, that if the cocoa crop failed due to blight or drought, then the plantation would make no money that year, and likely, the laborers would not be paid. On my walks I did meet two workers from a single plantation who said that they'd labored one year and nine months and had not yet been paid. Their names were Siaka Traore and Ibrahim Malé. They told me that the crops had failed the first year, and they were working a second because they hoped this season would be better, and they did not want to return home empty-handed.

I met the owner of this plantation. He confirmed, in a voice scarcely above a whisper, that his crops had died. Most of his teeth were missing. The roof of his home was half collapsed. His own infant son had a distended belly. During our brief talk, the infant sat beside us on the muddy ground, naked and wailing.

Another plantation owner, a man named Touré Fakourou, listened attentively to my translator as I spoke about the slavery accusations and the British video and the possibility of an international boycott of Ivorian cocoa. "We are not talking about slavery," he said, when I was finished. "We are talking about poverty."

Back in the city of Daloa, I tracked down a person who had worked as a translator for the *Slavery* documentary. His name was Michel Oulai, though he was better known by his pen name, Vincent Deh. He had written for a local newspaper called *Notre Voie*—Our Way—for nine years. He was thirty-six years old and spoke excellent English. I'd brought my copy of the documentary with me; Deh said he hadn't seen it. A friend of Deh's owned an old VCR and television, so we drove to the friend's house and watched the documentary together.

Deh, who appears in one of the scenes, seemed captivated by the movie, but when I told him that it was supposed to be accepted as truth, he considered this for a moment and said, "It's exaggerated." He did not seem particularly troubled by this notion. He assumed, he said, that this was how documentaries were made—for people to pay attention to your work, sometimes you had to exaggerate.

I told Deh that I was having trouble finding a single person who appeared to have been beaten. I asked him if, while working on the documentary, he'd seen the type of abuses that the film says are prevalent. "No," he said. "I never saw proof of even one slave." His comment about exaggerations, he told me, was just a polite way of saying the movie was false.

I asked him about one of the film's most powerful lines, in which a young laborer looks into the camera and says that people who eat chocolate are "eating my flesh." We rewound the tape, and Deh watched the scene again. He told me that the boy's statement couldn't have been genuine. Deh vividly recalled working with the laborer and said that the boy—like almost all the laborers—did not understand the relationship between picking cocoa and eating chocolate. Most kids, Deh said, didn't know what chocolate was. Deh felt that someone must have put the words into the boy's mouth, instructing him on what to say.

After speaking with Deh, I took my driver and translator and headed north, across the border and into Mali. The transformation was startling. Here, at the periphery of the Sahara, the soil seemed sapped and colorless; the corn was not even ankle-high by the Fourth of July, which I happened to spend in Mali. Beggars were everywhere. When I opened a tin of sardines, a crowd of children grabbed at me, pleading to drink the oil. I gave the tin to one boy, who was promptly pummeled by the others, the oil spilling onto the ground.

The first big city on the Malian side of the border is called Sikasso. The relief agency Save the Children Canada had recently opened a rehabilitation center in Sikasso to help treat the child slaves of West Africa for psychological problems. The facility was named *Horon So*—Freedom Center. It was located in a whitewashed stucco building, one of the nicest structures in town. Many of the younger boys who'd finished their time on the cocoa plantations were bused to Horon So from the Ivory Coast; some of these bus trips were paid for by the Malian Association of Daloa. The boys usually stayed for five days. They slept at the center, and ate there, and spoke with counselors. When I visited, Horon So's director was a thirty-five-year-old psychologist named Ibrahim Haidara. A few months later, he left Save the Children.

Haidara met with me one evening in the courtyard of my hotel. We ordered grilled chicken, drank a few Cokes, and embarked on a lengthy discussion. Haidara was born in France but was of Malian descent—his father was from Timbuktu, in central Mali. He spoke eloquently and openly and without any apparent agenda. He never asked me for money.

"I don't accept the word 'slave' to describe these kids," he said to me. "I have not seen any evidence of abuse from those coming back. Maybe a few machete marks, but nothing more. Almost all of these children want to go. They hang around bus stations, waiting for locateurs to take them across the border. For them, the Ivory Coast is a paradise."

Most boys, he explained, are accustomed to farm labor. They've worked on family farms, but for this work they are not paid. There are very few paying jobs available in Mali, Haidara said, but in the Ivory Coast there are jobs, so the boys cross the border. This has been the case, he said, ever since there was a border.

"Generally," Haidara said, "the children leave their home villages to get something they've wanted. They want what they don't have. The boys want bicycles, a radio, good clothes. They want basketball shoes. They know all the brand names. They want Nike basketball shoes. That is their dream." His job, Haidara explained, was to teach the workers the advantages of staying in Mali, with their families, and helping to improve the fortunes of their own country.

The next day, Haidara took me to meet Aly Diabate, the boy who'd been featured in the Knight Ridder series, the one whose story had been read aloud on the floor of the U.S. House of Representatives. Diabate had spent a year and a half on a plantation and a week at Horon So and was now back at his village, a scattering of huts along a brown river in the Malian hill country.

According to the Knight Ridder article, there were only "rare days" when Diabate wasn't flogged "with a bicycle chain or branches from a cacao tree." When I spoke with Diabate in person, the story was different. Perhaps because he was no longer beholden to the Malian Association, as he may have been when the Knight Ridder reporter spoke with him, he felt better able to speak freely. Whatever the reason, Diabate's tale had changed. He said that he did experience some physical abuse—he was slapped or hit with a fist by the brother of the plantation owner. This happened once or twice. But as for daily lashings, Diabate said this: "No, we were never whipped."

The Knight Ridder article reported that Diabate was fourteen years old. His extreme youth was a big part of the story. I visited Diabate less than a month after the article was published. I asked him his age. "I don't know how old I am," he said. My translator asked him to guess, and he said, "I guess nineteen." By this estimate, he would have been seventeen, not twelve, when he was hired to work on the plantation. It was difficult to tell Diabate's age by looking at him—he had the type of baby face that could allow an observer to believe he was far younger than he was, though his arms and shoulders had sprouted an adult musculature.

A day after speaking with Diabate, I returned to the Ivory Coast, back to the capital city of Abidjan. There, in a boxy high-rise, I visited a branch office of the United Nations Children's Fund, and met with an official named Lavender Degre. UNICEF had closely studied child labor in the region, and had issued numerous reports on the topic. Degre concurred, readily, that the slave story had been blown out of proportion. "We have never once used the word 'slave' in any of our reports," she said.

I asked her if the boys at the Malian Association of Daloa

would really lie to me, and she laughed. She wondered if I'd read any newspaper accounts about the so-called Street of Slaves. I had; the articles described a market in Abidjan where slaves could be bought and sold.

"If you give me a thousand francs"—about $1.40—"I'll give you a slave," said Degre. "It's all about economics. If you offer someone money for a slave, he will show you a slave. The street boys are smart. They'll go get their cousin and say, 'Look, here's a slave.' They'll get you all the slaves you want."

After three weeks in Africa, I realized I had my story. If you listened to certain members of the Malian Association and took notes and tilted your head just so—well, yes, there was slavery. West Africa is a very poor part of the world; if journalists were willing to pay good money to see slaves, it seemed as though some officials with the Malian Association were more than happy to provide them.

But I wanted to write about the real problem: I wanted to write about the crushing cycle of poverty, and about the suffering that young people were willing to endure in order to eke out a living. At the same time, I wanted to explain how the media can generate misunderstandings, and how aid agencies can perpetuate these errors. I wanted to demonstrate how we can sometimes see what we're looking for instead of what really exists. This wasn't the story I came to find, and it wasn't a particularly explosive one, but it felt important in its own quiet way. So I packed my belongings and flew home.

I described the idea to Ilena Silverman, my editor at the *New York Times Magazine*. I was excited about its prospects; it had the potential, I thought, to be an intelligent, insightful, unorthodox article. Silverman, though, said she wasn't particularly interested in yet another story accusing the media of getting everything wrong. She didn't want a piece that might unfairly harm humanitarian agencies. Instead, she suggested that I present all of these issues more palatably, perhaps by telling a detailed story of one boy. Weave an intimate portrait of a single laborer, she said, and through this one worker artfully clarify the fine line between slavery and poverty. "Could you do that?" she asked me.

I had spent almost all my time in Africa attempting to prove that the story I'd been sent to cover did not exist. But proving that there's no story, my editor had implied, is not itself much of a story. I realized that she was right. Her idea, the tale of one boy, seemed less complicated than mine, and possibly more profound.

Except that I'd just flown seven time zones from West Africa prepared to write one story, and now I was being asked to work on a very different one, using the same material. There was some part of me that knew, right then, that I could not fulfill my editor's request. I should have said so immediately. But I sensed that my success as a writer was almost solely in Silverman's hands, and I felt a powerful need to please her.

I couldn't even suggest a compromise, partway between her idea and mine—maybe a profile of three or four separate workers. That seemed to me like admitting I'd failed in West Africa. Also, I feared that any such compromise might result in a second-rate article, and one mediocre piece, I was convinced, was all it would take to damage my standing at the magazine and derail my ambitions.

So I began to rationalize. With all the interviews I'd done, many of them two and three hours long, I figured there had to be one boy who would work. Or, failing that, I could follow what Lavender Degre at UNICEF had said: If someone wants to see a slave, show her a slave.

Could I write a story about one boy? I told my editor I could.

SIX

WHAT I DID was take a handful of interviews and meld them together. One worker I'd spoken with told me about leaving his farm in Mali and traveling with a locateur to the Ivory Coast. Another described how he was sold to a plantation owner. A third detailed the type of labor he did on his plantation. And yet another spoke of his time with the Malian Association of Daloa and Save the Children, and of his return home. I lifted details and quotes from all these stories, and a few others, and invented a single character who narrated the entire journey as if it were his own.

I thought I'd get away with it. I was writing about impoverished, illiterate teenagers in the jungles of West Africa. Who would be able to determine that my main character didn't exist? For several months after the article was printed, it seemed as if my instincts were right. I had gotten away with it.

Then I was caught. My career, twelve years of intensely focused labor, promptly imploded. I was about to be pilloried on page A-3 of the *New York Times*. The rest of the journalism world would soon weigh in. I'd be shown, publicly, to be a liar—a stink you can never fully wash off. I planned to go into hibernation. And then my phone rang.

On the other end of the line was Matt Sabo of the *Oregonian,* asking to speak with Michael Finkel of the *New York Times*. When he said that, I winced. I had spent all of my adult life trying to become Michael Finkel of the *New York Times*. Now, after scarcely a year, I was finished.

At first, our conversation was confusing. The *Oregonian* reporter said that he was calling because he was writing a story about Christian Longo. Until that instant, I had never heard of Christian Longo. The reporter told me that he was working on a lengthy piece about the crimes Longo had been accused of, about his run from the law, about the details of his capture.

"But why are you calling me?" I asked.

The reporter explained. After Longo escaped to Mexico, he changed his identity, which is not a surprising action for a most-wanted fugitive. But rather than creating a fictitious alias, he took on a real one. And apparently he had done an excellent job convincing others of his new persona.

While Christian Longo was in Mexico, wanted for the murder of his wife and three young children, he pretended to be a journalist. He chatted with other tourists about the stories he had written; he said he was in the Cancún area on assignment. He took notes. He teamed up with a photographer. And his name and newspaper, he told many of the people he met, was Michael Finkel of the *New York Times*.

PART TWO

MEXICO
MEXICO

SEVEN

BY THE TIME the *Oregonian* reporter called, I had already exiled myself to the upper floor of my home, a few miles outside the town of Bozeman, Montana. My hibernation had not officially begun—I was still awaiting publication of the Editors' Note—but there seemed no place else to go. I had squandered my career due to stupidity and hubris; I had caused my own downfall. I did not want to see my friends or speak with my parents. I felt remorseful and ashamed and confused. I don't know what I wanted, except to blame someone else for my deceit.

Hours at a stretch, I lay prone on the upstairs sofa, burrowed beneath my laundry pile. Or else I paced back and forth in my bedroom. When my head began to pound—when I was so furious at myself that my vision went fuzzy—I'd clamp my palms over my ears and yell at the ceiling until my breath gave out. All day, I wore sweatpants and bedroom slippers. I didn't watch TV or listen to music. I ate whatever canned foods were left in the house. More than once, I crawled into the cramped, dusty space underneath my writing desk and tore at the carpet, rubbing my fingers raw.

It was at this point that the reporter phoned. The story he told me was so absurd and unexpected, and delivered with such impeccable timing, that it slapped me from my brooding. All at once, I was curious and repulsed and perplexed. And then my immediate feelings coalesced into one distinct, uncontainable reaction: I laughed. I really did. Out loud, over the phone, to the reporter from the *Oregonian*.

The Editors' Note was printed the next morning. I remained in hiding for a spell, but I felt as though I could breathe again. I'd been released from my loop of self-centered moping. I ventured to the supermarket for supplies; I rented a few movies; I peeked on the internet to learn about Longo and read of my disgrace. As the other media outlets weighed in—one journalist compared my ethics to those of a "glazy-eyed person" who kills abortion doctors—I remained passive and distant, saddened in a stunned sort of way, as if watching my belongings consumed by a fire. I took my beatings, and then, once the story had played itself out, I picked up the phone and called the *Oregonian* reporter.

I had only one question: How could I get in touch with Christian Longo? That was impossible, the reporter said. Longo's lawyers—he was represented by two public defenders—had forbidden their client from speaking with the press. Nearly every West Coast news outlet from Seattle to San Francisco had requested an interview, and not one, the reporter told me, had been accommodated. Even so, on March 6, 2002, two weeks after the Editors' Note had appeared, I wrote Longo a letter. I filled the front and back of one sheet of yellow lined paper.

Here, in its entirety, is what I wrote:

Dear Mr. Longo:

Yes, it is actually me—Michael Finkel of The New York Times. Or, rather, formerly of The New York Times. To tell you the truth, I was just recently fired. I invented a character in one of my recent stories, and I was caught, and was very publicly fired. So now I am out of a job. This is why I am writing this by hand rather than computer—I'm actually no longer an official journalist, though I still love to write.

I understand that while you were in Mexico you used my name. I do not mind this at all—in fact, I find it both interesting and, in a way, it makes me feel somewhat honored. I understand that you are facing an upcoming trial, and that there is probably much that you are unable to talk about, but I was hoping that you would agree to meet with me in person.

I live here in Montana, which is not much of a long drive away. I'd like to ask you why you chose to be "me," and what it felt like, and maybe talk with you a bit about this. We can even talk about writing, if you want.

I'd like to do this because at the same time that you were using my name, I lost my own—my firing, as I mentioned, was very public. During my firing, I was robbed of the two things that a freelance writer needs to survive—his name and his reputation. Both are now gone.

Now that I'm out of a job, I am sort of seeking to find out who I really am, and I would be grateful and honored if you would consider speaking with me. Please write me back—my address is on the front of this note—or call me collect. Please let me know when you are willing to meet, and I will be there.

I look forward to hearing from you soon.

Yours,
Mike Finkel

I photocopied the letter, then mailed the original to Christian Longo, care of the Lincoln County Jail.

EIGHT

A MONTH PASSED. There was no reply. Except for his lawyers, it seemed that Longo was not speaking with anyone. In the Lincoln County Jail, he was being held under administrative-segregation status, which meant that he was alone in his eight-foot-by-eleven-foot cell, alone for all of his meals, alone during his time in the recreation yard, and forbidden from communicating with all other inmates. The window in his cell was frosted over, denying him an outside view. He was completely sealed off. I didn't know if he'd even received my letter.

As for me, that month was particularly uncomfortable. Within days of publishing the Editors' Note, the *Times* announced that a thorough investigation would be made into all the stories I'd written for the paper. Reporters were mobilized in Haiti and Israel and Afghanistan to reinterview people who'd appeared in my articles. I didn't blame the *Times* for looking into my pieces. In other cases of journalistic fraud—most notably in the late 1990s, with Stephen Glass, who wrote primarily for the *New Republic,* but also fifteen months after my incident, with the *Times* reporter Jayson Blair, and even more recently with *USA Today* writer Jack Kelley—where one deceitful story turned up, soon there were many.

I knew in my case there was indeed only one. And if that were shown to be so, I hoped to be able to return to journalism. But I couldn't be sure what the investigation would find. I didn't know if I had been tricked on an assignment by a manipulative source, or if a person I'd spoken with would significantly change his or her story upon a follow-up interview. If the *Times* had even a hint of suspicion, I sensed I'd have no recourse. I had been caught lying once, and therefore was unlikely to be further believed. I had no idea what I'd do if my journalism career was over.

The silence from Longo, the ongoing investigation into my articles, and the meaningless, meandering days I wasted cooped inside my home started to wear on me. Over the previous decade, I'd spent at least six months of each year on the road, accumulating the raw materials for magazine articles. My time at home was devoted to writing, to researching, and to skiing or hiking in the mountains. I wasn't accustomed to stasis. Before long, I was again pacing my bedroom. The excitement that the *Oregonian* call had stirred in me faded away, and my brooding resurfaced. I felt an acute need to escape. So I climbed in my pickup truck and left town.

I drove through southern Montana and into Wyoming. Spring was approaching; it was warm enough to roll down my windows. I drove through Colorado and Kansas. I slept in cheap motels, or I drank coffee and drove all night. For five hundred miles, I tortured myself by thinking of ways I could have written the West Africa story without cheating. For another five hundred, I berated myself for not simply having told my editor the truth.

I drove through Oklahoma and across Texas. I began calling people on my cell phone. I spoke with my parents, my sister, and my friends. Some of the things they said about the *Times* disaster were surprising. My mom told me that, in many ways, she was relieved I'd been fired. She said I'd gotten myself into a crazy cycle with the *Times*. Every story I wrote, it seemed, had to be bigger, and better, and more daring. My friend Doug called it "a giant game of 'Top This.'" Another friend candidly informed me that I'd become an asshole; that writing for the *Times* had made me frenzied and rude and cocky.

Mandi described me as overworked, underslept, and over-whelmed. Mark said I was utterly career-obsessed. "Some people live and learn," he told me. "You just live." My sister said I would never have quit the *Times* job on my own, and that she had feared, as my parents had, that my intensity and my risk-taking may have ended up costing my life.

One friend told me that she cared less about the *Times* incident than the insensitive way that I'd treated her. For a few weeks, she said—this was while I was still working for the *Times*—I'd fawned over her as if I were interested in initiating a serious romance, and then, without warning, I'd dropped almost completely out of contact. Another woman told me the very same thing. "It was like you were pretending to be some sort of character," she said. "You can stop that now and start being an actual person."

I drove across New Mexico and into Arizona. I thought about what my friends and family had said. I realized that what I'd done with the *Times* article wasn't entirely a random event. There'd been an inflation of ego, a buildup of stress, a ratcheting of risky behavior. By the time I wrote the West Africa story, I'd become so manic and arrogant that I assumed the rules of journalism no longer applied to me.

I was still driving when I learned that the *Times* investigation into my other stories had been completed. There was nothing to report but a single spelling mistake and an inconsequential numerical error. This was gratifying to hear, though it didn't much boost my spirits. I was still ashamed of myself.

Five thousand miles into my road trip, midway through Utah, I spoke with my friend Mark again. This time, he called me. He'd been taking care of my house, and he knew about the letter I'd written to Christian Longo. "There's a strange phone number on the caller ID," he said. Mark hadn't been there to pick up the phone, but he figured I'd want to know. The number itself wasn't strange, though the area code, 541, was unfamiliar. But beneath the number, Mark said, where the caller's name or business is usually displayed, it read this: INMATE PHONE.

Immediately, some of my gloom lifted. I programmed my home telephone to forward calls to my cell, in hopes that Longo would try to contact me again. I was already on my way back to Montana, and I filled the miles by imagining what kind of conversation we might have. I found myself flush, once more, with enthusiasm. A few days later, on Tuesday, April 9, 2002, while I was parked at a desert overlook near the town of Moab, Utah, the phone rang and it was him.

NINE

YOU CAN'T SPEAK with an inmate of the Lincoln County Jail right away. The first thing you hear, when you receive a collect call from the jail, is a recorded message. It's in a woman's voice. "To refuse this call," the message starts, "hang up." Then it says that the call may be monitored or recorded, and mentions information on "terms and conditions" and "binding arbitration" and "limitation of liability." This goes on for about a minute. Finally, the message says, "To accept this call, dial one after the tone. Please make your selection now. Thank you." Then there's a beep.

I dialed one. The phone lines opened.

"Mr. Longo?" I said.

There was no salutation, no small talk. Christian Longo greeted me with a question. "How do I know," he asked, "that this is the real Michael Finkel?"

I was taken by surprise. In every conversational scenario I'd envisioned, I was the one asking questions. I was in charge of the discussion. Yet now, in the course of a single exchange, Longo seemed to have grabbed command, and I was the one scrambling to reply.

"I'm not really sure how to answer," I said. "If you were here, I could show you my driver's license. I don't suppose you know my Social Security number or my mother's maiden name?"

"I don't," said Longo.

"Well," I said, "I think you're going to have to take it on faith."

"Not good enough," he said. "Any journalist could've written me that letter, trying for a scoop." He said this in a friendly way, though with a hint of challenge, as if he were prodding me to think through a riddle.

But I was stuck. "I don't know what to tell you," I said.

"I was prepared for that," he said. Beyond the audioscape of his flat midwestern vowels, I envisioned a brief, smug smile. "I have a couple of questions for you."

"Great," I said, relieved and somewhat amazed. I was clearly not the only one who'd spent time thinking about our talk. "Go ahead," I said.

He paused for a moment, and his voice shifted into a less collo-quial cadence, as if he were reading. He was, as I later found out: He'd prepared a test for me, thirteen questions in all, complete with answer key. "Okay," he said. "What's the name of the main character in the story that got you fired?"

Good question, I thought—the type of question a fake Michael Finkel probably wouldn't know. How Longo himself knew this was unclear, though I figured I'd learn soon enough. First I had to ace his test. "His name is Youssouf Malé," I said. Then I added, going for extra credit, "The last name is spelled m-a-l-e, though it's pronounced ma-lee. There's an accent on the end. An accent aigu." I was pleased with the thoroughness of my answer. "So," I asked, "does that prove I'm me?"

"No," he said. He continued reading from his test: "In that story, how much money did you say Youssouf Malé earned in a year?"

"Tough one," I said. From the time I'd written the article until the time Longo asked me this question, more than eight months had passed. "Details like that don't tend to stay with me. I think it was a hundred and fifty dollars."

"One hundred and two dollars," Longo said. He sounded skeptical.

"Give me another," I said.

"What was the headline of the last story you wrote for the *New York Times Magazine*?"

I couldn't believe it. I'd actually written two articles in the time between the publication of my child-slave story and the uncovering of my deception. Both of the pieces were about the war in Afghanistan; the second, printed only days before I was caught, was about a group of villages trapped between the Northern Alliance and Taliban fronts. The problem was that I didn't write the headline. I never wrote headlines for my *Times* articles; the editors did. And with all the trauma of the firing, I'd scarcely glanced at the issue in which my last story appeared.

"It's funny you should ask this," I said, attempting to divert the question with a touch of levity. I told him that writers don't write headlines. I said I could nearly picture the headline in my mind. "It was five words long," I said. "It was 'To Stay or to Go,' but not quite that. 'To Hide or to Seek.' No. 'To Something or to Something Else.'"

"'To Wait or to Flee,'" Longo said, pointedly. I got the impression that he felt he'd outfoxed an impostor. I couldn't blame him. A tightness came to my throat; the first stage of distress. It was possible that I'd lose my chance to speak with Christian Longo because I was unable to prove that I was really me. Somehow, it seemed a fitting punishment.

"Hold on," I said. "I know all about that story. I can probably recite the opening paragraph."

I happen to be a slow, methodical writer, and every hour or two while I'm working, I tend to reread my manuscript from the top, so the beginnings are usually burned into my memory. "Here's the first sentence," I said. "'They had a radio, just a single battery-powered radio, so the news traveled by word of mouth up and down the footpaths of Abdulgan, village to village, until everyone knew.' That's the opener. The second sentence is, 'They knew what was happening elsewhere in Afghanistan.' The third is—"

"Okay," said Longo, cutting me off. "I believe you."

TEN

AND SO WE were free to talk. For a long moment, though, there was only silence. Longo was clearly waiting for me to say something, but I was unsure how to begin. He'd made no public statements since his arrest three months before, and I felt it was important that his first impressions of me put him at least somewhat at ease. But what do you say to a man who has likely murdered his family, then fled the country and stolen your identity?

"Call me Mike," I said.

"Call me Chris," he said.

I asked him why he decided to phone me. He said he'd read my letter several times, and had debated making contact. My letter, he told me, was the first he had heard of my firing, and he'd asked his lawyers to verify its truth. The lawyers brought him a copy of the Editors' Note, the child-slave story, and my Afghanistan articles, which was how he'd been able to quiz me so thoroughly.

"In your letter," Chris said, "you wrote that you weren't upset with me for using your name. But the more I thought about it, the more I felt responsible for you losing your job. It just seemed too much of a coincidence, and I wanted to know if I was in any way to blame."

He was concerned, he explained, that his actions in Mexico or the publicity surrounding his arrest had somehow exacerbated my troubles. He said this worry was so great that he felt compelled to call. He added that I was the only person he'd phoned since he had been incarcerated—he hadn't spoken with friends or family, not even his parents, and certainly not with members of the media. His lawyers, he added, did not know about this call, and if they found out, they'd likely throw a fit.

It seemed clear, from the way Longo had made me submit to his quiz, that he was a cautious man. I found it odd, therefore, that he would ignore the counsel of his legal team merely to learn if he had damaged my writing career. But if there was another reason he'd contacted me, this didn't seem the right moment to pry.

I assured Longo that he had nothing to do with the *Times* disaster. I'd written my fake story, I pointed out, long before anything happened to his family. That's actually the way I phrased it: "before anything happened to your family." I was careful not to say "before you murdered your family" or something similar. There was no need for bluntness. Longo had yet to enter a plea to the charges; therefore, at this moment, he was legally innocent. And though the facts of the case did not look good—four dead bodies found in Oregon, one live man found in Mexico—I had to concede it was possible that Longo was actually innocent. So rather than speak to him as a person who had committed a terrible crime, I addressed him as a person to whom something terrible had occurred.

"Why," I asked, keeping the conversation on safe ground, "did you decide to impersonate me, of all people?" I had actually developed a theory about this. I had assumed that while Longo was escaping from Oregon, he'd somehow come across a Sunday *New York Times*. Many people keep sections of the Sunday paper lying around all week. I'd had an article in the *Times Magazine* on December 16, 2001, the day before the murders were thought to have occurred. My byline, Michael Finkel, was printed on the magazine's cover. It's a rhymy and rather funny name, and therefore perhaps easy to remember. (When I was young, kids would tease me by playing "Michael Finkel" instead of "Marco Polo" in the local swimming pool.) Longo, I figured, had spotted my name and borrowed it as his own—a random act.

My theory was wrong. Longo told me that he'd long been familiar with my work, and not just from the *Times*. He'd read my stories in *Skiing*, and *Sports Illustrated*, and *National Geographic Adventure*. He said my articles appealed to him. He'd always thought, he said, that if he were to become a journalist, he'd want to write the same sort of stories that I wrote. He knew so much about my articles, he added, that he'd been able to speak about them, confidently and convincingly, while in Mexico. He explained all this in a droll, relaxed manner. "You have a writing style," he said, "that I wasn't embarrassed to call my own."

In other words, Longo was a fan. And there is perhaps nothing more dangerous to a writer's common sense than encountering an enthusiastic reader of his work, even if he's calling collect from county jail. During our conversation, I jotted quotes and impressions in a notebook, and as Longo continued to praise my work, my objectivity began to soften. "A v. nice guy," I wrote down.

I sustained the patter by asking the most basic, blind-date sort of questions, then exclaiming eagerly about any similarity I uncovered. For example: We both had January birthdays! Longo had recently turned twenty-eight, two weeks after I'd turned thirty-three. Neither of us was a native of the West—he'd grown up in suburban Indianapolis; I was from suburban Connecticut. He'd been a Jehovah's Witness but had been kicked out of the organization. I was a lapsed Jew.

He told me that he felt battered by the media's coverage of his case. "There's no way you can know me from reading the papers," he said. I told him I understood exactly what he meant. He said he had never written anything for publication but had once worked for a company that handled home delivery of the *New York Times*. "I was always proud to say I worked for the *Times*," he told me. I was always proud of that too, I said.

It was clear that Longo wanted to keep the conversation light—he chuckled at even the slightest trace of humor, releasing a quick, staccato "heh-heh-heh-heh." Of course, I wanted to ask about the murders. But it wasn't the appropriate time. So instead of murder we spoke about travel, and skiing, and the flavoring we preferred in our lattes (me: vanilla; him: Irish Cream). His voice was as controlled and steady as his laugh; no shouts, no whispers, scarcely any inflection. It was a voice that could be transcribed into text without ever needing an exclamation point. He was partial to repeating the word "gotcha" as a conversational space-filler.

The discussion flowed with no uncomfortable silences, though the whole thing—a casual, bantering chat with a man who'd recently been a Most Wanted fugitive—felt more than a little surreal. Once it was over, I was wildly energized, as if I'd been freed from some confinement, and I had to put on my running shoes and go for a jog to settle myself down. When I returned to my truck I sat in the driver's seat, panting, and spoke my thoughts into my pocket-sized tape recorder.

Everything Longo said had seemed honest, until late in our talk, when he mentioned his time in Mexico. "If I hadn't been caught down there," he told me, "I was going to fly home anyway, and turn myself in." This may or may not have been true—there's no way to know—but to me it sounded like a lie. Who would swap a beach resort in Mexico for a jail cell in Oregon?

Our conversation had ended abruptly. When we'd spoken on the phone for nearly an hour, there was a loud beep on the line. Longo told me that this was the jail's indication that we had only a few seconds remaining before the line was cut. I took this moment to ask him if I could come for a visit.

"Let me just check my schedule," Longo responded, dryly. "Well, yes, I think I might be able to find the time."

ELEVEN

THE LINCOLN COUNTY JAIL is a shoebox-shaped building, made of cinder block and brick, solid-looking on the outside save for two rows of slits, like dashed lines, that mark the cells' windows. When I arrived at the jail's entrance area, an officer instructed me to remove my belt and leave it in a locker, along with my car keys. I passed through a metal detector and then was directed into an oversize elevator that had stainless-steel doors and a rubberized floor. There were no buttons to press. The doors shut; the elevator rose; the doors opened.

The jail's visiting room consisted of five booths. Each had a short metal stool bolted to the floor in front of a gray metal shelf, mounted at desk height. A black telephone receiver hung on the booth's left-hand side. Embedded in the wall over the shelf was a thick square of reinforced glass, through which I could see, on the prisoners' side, another stool, shelf, and telephone. Every booth was empty. Longo's segregation status in the jail meant that no other inmates could be present while he was in the visiting area.

I sat at the far booth and waited. I studied the fingerprints on the window, some of them tiny, clearly children's, and I counted the kiss marks. I dried my palms on my pants. Graffiti had been scratched into the desk: "Bobby R"; "Tammy"; "Hi Dad"; "I love you so much Chava"; "Amanda y Edgar"; "Eat Shit"; "Fuck Toad." The scent in the air brought to mind fresh paint and sour milk.

Longo strode into the room, energetically, as if he were ready to sell me something. He was wearing a navy blue sweatshirt that read LONGO in black permanent marker on the front, blue sweatpants that read JAIL in white silk-screened letters down the right leg, and tan plastic sandals and white socks. He carried a large brown envelope, a yellow legal pad, and a golf pencil outfitted with an arrow-shaped eraser. He was on the tall side—a shade over six feet, he later said—and fit-looking. He sat on the stool across from me, and we peered at each other through the smudged window.

He had a baby face—that was my first thought—with a scattering of freckles across his cheeks and not so much as the hint of a beard. His ears angled sharply outward; his hair, short and neat, was either reddish blond or blondish red. He had hazel eyes, long eyelashes, pale skin, and an utterly characterless nose, the kind of nose that people who get nose jobs always want to have. His features had achieved an enviable harmony—he was casually good-looking, in a sporty, fraternity brother sort of way, and he managed to appear self-assured even in his jail uniform.

There was nothing about him that seemed remotely scary, and he was sealed off behind a slab of bulletproof glass, with armed guards watching from their own glass-walled station inside the jail. But despite all this I felt a twinge of genuine, stomach-tightening fear. The person facing me was considered so dangerous he was not allowed near other criminals—not even permitted, or so it appeared, to share the same *air* as anyone else, as if he were the carrier of some lethal disease.

He smiled at me briefly, just the top teeth showing, and we each picked up our phone. We swapped hellos and how-are-yous, and in the midst of our greetings, while maintaining eye contact, I flipped open my steno book and scribbled a few notes. I couldn't help myself. I'd lost my job, not my instincts. Longo's story—one that combined murder, identity theft, and a bizarre personal connection—was the journalistic equivalent of a winning lottery ticket.

Pursuing such a story was irresistible to me. In fact, from the moment the *Oregonian* reporter had called, I'd had a vague sense that the beginnings of my redemption, both professional and personal, might somehow lie with Longo. His tale could provide me with a chance to return to journalism. And I thought that if I were able to be truthful with Longo—an accused murderer and a possible con man; a person who might easily forgive deceit—then I'd demonstrate, at least to myself, that I had moved beyond the dishonest behavior that had cost me my job. On top of all this was a morbid but undeniable curiosity: If Longo was indeed guilty, I wanted to know what could possibly drive a man to murder his own family.

So I'd brought along a pen and notebook to the Lincoln County Jail, and I began taking notes. My pen's movement caught Longo's attention, and he looked through the glass wall between us, down at my hand, and said, "What are you writing?" He had a concerned expression on his face.

I may have decided that working on a story about Longo was an ideal way to begin repairing my life, but it dawned on me, just then, that Longo himself might not be willing to cooperate. I'd thought he would. During our initial phone call, shortly before this visit, Longo had said that he was distressed by the media's coverage of his case; though he hadn't spoken with any reporters, he'd repeatedly been portrayed as a conniving, psychopathic killer. Longo had also revealed, over the phone, that he admired my writing. I had responded that maybe I should be the one to tell his story, perhaps in a magazine article. Longo had seemed receptive to the idea. "There's a lot of things that haven't been said," he'd told me, in an agreeable tone of voice. This was a chief reason I'd made the trip to Newport—so that Longo could begin saying what hadn't been said.

But now, sitting in the visitor's booth and sensing his displeasure, I wondered if he had changed his mind. Perhaps he'd realized that, in light of my firing, my usefulness as a mouthpiece was minimal. Possibly, he had understood or been warned that speaking with a journalist, even a defrocked one, could only hurt him when it came time for his trial, which at this time, early May of 2002, was at least six months away. Maybe he'd agreed to meet me for no other reason than to break up the monotony of jail and see what the person he'd masqueraded as really looked like.

While I paused to consider whether I wanted to tell Longo what I'd written in my notebook—not that "young-looking, ears stick out" is such a revelation, but I was caught off guard and didn't want to make a blunder—he spoke up and answered his own question. "You're probably writing your first impressions," he said. I confessed I was. "I'd be doing that, too," he added knowingly, as if he were also a journalist, swapping tricks of the trade with a colleague.

I displayed my notebook so that he could see what I'd jotted down, and for the remainder of the thirty-minute visit I took few notes. Immediately after, I would hurry to my hotel room and record every detail I could remember.

Our conversation had a peculiar momentum. We'd be discussing the blandest of subjects—which local restaurants I'd eaten at, what hotel I was staying in—when some word or phrase seemed to generate in Longo an intense emotion, and he'd appear on the verge of revealing an intimate thought before he would regain his composure, hurriedly change the subject, and settle back into blandness. When I mentioned, for example, that I'd taken a walk along Newport's bayfront, it was the word "bay" that sparked a reaction.

"I'll never look at the bay again," Longo told me. At first I thought he was lamenting the fact that he might spend the rest of his life locked in prison. But that's not what he'd implied. Longo added that he was thankful his cell's window was frosted over. "That way," he explained, "I can't see the water."

I grasped his meaning. "Those were the waters," I said, carefully maintaining a nonaccusatory tone, "in which your family was found."

He nodded yes, and I looked at Longo. I stared for a couple of beats too long. But his eyes revealed nothing. He returned the stare and said, "I just wanted to do the best I could for my family," and his eyes moistened and he glanced away and I thought he might weep. But when he looked back, he seemed fine. "You know," he said, "I was born in a town on the Mississippi River."

That's how it went: Longo speaking through a scratchy phone, bouncing his pencil with one hand, holding the receiver with the other (his fingernails unbitten and dirtless), telling me about his birthplace of Burlington, Iowa, and his great-grandparents' pig farm (his Adam's apple, small and sharp, floating up and down like a buoy), then mentioning a song that played while he was alone in the jail's rec yard—"Hero" by Enrique Iglesias—and how it made him ache for his family ("It reminded me of how I should have been"), and me nodding sympathetically and saying, "I understand, I understand," but all the while thinking to myself, "Let's get on with this, let's talk about the murders."

It seemed as if he just wanted to chat as normally as possible, face-to-face—or, as he said, "face-to-glass-to-face." Longo, I felt, had no interest in answering interview-style questions. He didn't want to play the part of Chris Longo, accused family killer. He wanted to be Chris Longo, above-average Joe.

He had even come up with a mathematical technique that demonstrated precisely how ordinary he was. Longo did not explain the computations to me, but during our visit he did share the results: He had been a decent, regular guy for 92.88 percent of his life. That's what he said. The specifics of the remaining 7 percent and change were left undiscussed.

When our thirty minutes were nearly over, Longo held up the brown envelope he'd carried into the visiting room. My name was penciled on the front. "I've written you a letter," he explained. And then, as if all of this—the call, the visit—had been some sort of entrance exam, an odd type of tryout, he said, "I'm going to decide whether to mail this to you or not."

TWELVE

HE DECIDED YES. The letter came in the same envelope Longo had shown me. It had surprising heft. Inside was a stack of yellow paper with faded blue lines; every page was covered, top to bottom, left to right, in immaculate penciled print, the letters grammar-school tidy, each line a calm string of boxcars. There were no signs of erasure, almost no scratch-outs—it was as though his thoughts had flowed from head to hand in a boulderless stream. And a hell of a stream it was: He had written seventy-eight pages, all with a golf pencil, the only writing instrument he was permitted to use. It was the longest letter I had ever received.

"Dear Mike," he began, and then, after a brief preamble, he opened into a rant: "I sometimes feel like a caged animal. I know that I can speak, I do have a voice, but the guards look at me as though I'm speaking with the language of an ape; the words hit their faces & fall to the floor, w/out expression. They're so officious; any vestige of free will is lost."

Longo clearly had a lot to say and badly needed someone to listen. It was obvious, too, that he felt far more comfortable writing than talking—he hadn't previously uttered a single ill word about the guards or his treatment. This made sense. Phone calls and visits were likely to be monitored. My letters to Longo, as with all letters sent to the Lincoln County Jail, were opened and inspected before delivery. But Longo's letter to me, if the jail followed its customary procedures, should have been sent untouched.

There was no law requiring inmates' letters to be mailed unopened, but according to Longo, unless there was a suspicion of contraband being sent, the jail rarely examined outgoing mail. Longo had devised a plan to test whether the jail was following its usual routine with him. Soon after I received the letter, Longo phoned me again. He'd carefully sealed the envelope himself, he said, and instructed me to inspect the flap for signs of tampering or regluing. I saw none. He asked that I examine the writing on the front to make sure it was his. I said it appeared to be. Finally, he told me to rip the envelope in half and check the inside. He wanted to know if there was something written in there. I looked closely and saw, in minute print, the word "fire."

"Good," Longo said. Before he'd licked the envelope, he had reached his hand inside and penciled the word. He was now convinced that jail personnel had not opened the letter and then resealed it, nor looked at the letter and then mailed it in a new envelope. We at least had a one-way line of secure communication.

I hadn't finished the first page of Longo's letter before I saw more clearly why he'd impersonated a journalist. After all, he could have pretended to be anyone—a stunt man, a soldier, an emergency-room physician. But his selection, I saw, was at least in some ways logical. In his letter, he wasn't merely imparting information: He was trying to *write*. His sentences were often rhythmic and complex; he experimented with metaphors; he was willing to dip into his stockpile of vocabulary words ("officious," "vestige").

One thing Longo did have trouble with was spelling. As I read his letter, I came across the words "definately" and "rediculous" and "abnoxious," along with a spate of incorrectly placed or erroneously omitted apostrophes. When quoting from his letters, and mine as well, I've preserved the content exactly as composed, incorrect spelling and grammar included.

The fourth page of his letter opened with a headline that read "First Impressions." This section was obviously inserted after our visit, for it mimicked the moment in our meeting when I'd recorded my initial thoughts about him. "Stereotypical journalist" was his opening entry. Then: "Rectangular glasses, thinning hair, intense stare looking deeper into my own eyes trying to see if truth is at the surface or somewhere deeper. Slight man. Mix of tall jockey meets chess club." Longo, it seemed, was trying to demonstrate that he was not in the least intimidated by me. His description also made me feel unnervingly exposed, as if he had the ability to peek into my thoughts.

The letter's next paragraph, under the heading "Biggest Fear," began with this: "You are a typical journalist out to get a story by whatever seedy means possible." He continued by reminding me that there was no shortage of reporters who wished to interview him, and that if he so desired he could, as he put it, "leap higher" than me. Both ABC News and NBC News, he claimed, had offered him a chance to appear on prime-time television; *Time* magazine, he said, wanted to put him on the cover. Why should he tell his story to a disgraced journalist, he seemed to be implying, when plenty of respectable ones were willing to listen?

I suppose this was his method of asserting who had the real authority in our relationship. Longo had dozens of reporters pursuing him, while I currently had no meaningful story to tell except his. The way I read it, if Longo spoke with me, it would be an act of charity; a form of pity, even. And the person with the power, of course, establishes the rules. Which Longo did. "You're going to have to be completely open & honest w/ me," he wrote. No games, no bull. "If you want the scoop," he continued, "tell me."

Though it seemed he didn't need to be told. The request was rhetorical; it appeared on the fourth page of his letter, right after his "Biggest Fear" section. He was well aware that I was interested in writing about him, and apparently he just needed to make sure I understood our power structure.

This accomplished, he proceeded to launch into the heart of the letter—the story of his time on the lam, in Mexico, during the weeks after his family was murdered. He titled this section the "Michael Finkel Affair." It opened with a description of an airplane ride on the morning of December 27, 2001, as Longo was preparing to touch down at the Mexican beach resort of Cancún. The writing continued for more than seventy pages, his paragraphs swollen with memories and details and tidbits of conversation. It was, beyond all expectations, an opening to the story I was seeking.

THIRTEEN

OVER THE CARIBBEAN came the plane, low and smooth, the tourists at the windows craning their necks, viewing what Longo described as "the gem-like blues of the sea." Longo was sitting on the aisle, alone in his row, and the vacant seat next to him triggered the thought that his wife, MaryJane, should be with him. "A mountain of guilt avalanched on top of me," he wrote. "I couldn't stop thinking how she would like nothing more than to be seated there, in that empty seat, w/ that giddy smile on her face."

The plane landed. Longo retrieved his small brown backpack from the overhead compartment and stood in line at Immigration Control. When he reached the booth, Longo wrote, he flashed the officer his most self-effacing smile. The officer looked at his birth certificate—Longo didn't have a passport—and then, without asking a single question or typing one word into his computer, stamped a tourist card and admitted him into Mexico.

Outside the airport, he boarded a minivan bound for Cancún-area lodgings. He had less than $200 with him, so he was going to have to live cheaply, and perhaps find a job once his money ran out. When he'd booked his ticket to Mexico, over the internet, using a stolen credit-card number, he'd also read up on local hostels, so he had a general idea of where he wanted to stay.

Eight other tourists, chatty and friendly, boarded the minivan.

Longo hadn't spoken with a single person since leaving San Francisco, but now, in the minivan, he realized he couldn't maintain his silence much longer. As the van departed the airport, he was bombarded with questions. What's your name? Where are you from? How long are you staying? One elderly woman, noticing his wedding band, said, "Oh, you're married? Do you have any kids? Are they here, too?"

Longo felt obliged to answer. "No," he responded to the elderly woman. "They've left me for now, so I'm on my own for a bit of a vacation." Tears began to well, and the woman winced and apologized to him, and he was able to avoid divulging anything further for the remainder of the ride. But he realized he couldn't go around weeping every time someone asked him a question. "I decided at that moment," he wrote, "that nothing in my past could be a reality here." He switched his wedding band from his left hand to his right.

He stayed downtown, away from the beachfront resorts, at a basic place with a generic name: Mexico Hostel. Four to a room, bathroom down the hall, ten bucks a night. He stashed his luggage in a locker at the hostel and walked to a grocery store and purchased bread, milk, eggs, an English-language newspaper, and a six-pack of Dos Equis. He cooked an omelet in the hostel's communal kitchen, then caught a city bus headed to the beach.

Stretched out in a lounge chair on the beach, he thought for the first time about constructing a plausible alias. Little came to mind. Instead, he read the newspaper he'd purchased. He scanned the classifieds for a possible employment opportunity. Nothing seemed promising. Then he turned to the travel pages. Skimming these articles, he wrote, reminded him of his favorite Sunday-morning ritual, when he would go to the local Starbucks with his family and order coffee for him, tea for MaryJane, and hot chocolate for the kids, and read the *New York Times*. His favorite parts were the travel section and the magazine.

And there, on the beach, an idea came to mind: "A perfect facade," as he put it. For years, it had been a fantasy of his to become a professional adventurer. Now, it occurred to him, was the ideal opportunity. "I could live out a dream," he wrote. And why not adopt the guise of a writer he'd often admired—the one with the rhyming name, the one whose trips often struck what he called his "jealosy bone"? Longo's middle name is Michael. Was it really such a big leap to become Michael Finkel? No, he thought, it wasn't.

So that part was settled: the outer part, the Michael Finkel layer. What wasn't settled was everything beneath—the Christian Longo part. That night, his first in Mexico, he found himself unable to sleep. The hostel was filled with vacationers, most of them young and sociable, and the atmosphere of endless revelry, he wrote, clashed horribly with his remorse about the decisions he'd made in the past few months. He realized that his relatives and friends were likely frantic with grief over the news of his family. Lying in his bunk, more than two thousand miles from Oregon, he felt an intense loneliness, and the familiar burn of tears.

This tug-of-war—between the frivolity of Cancún and the weight of his distress—continued for the whole of his trip. "Every day seemed the same," he wrote. "When I was by myself, either trying to rest, or wandering around mindless through the city streets, I'd find myself feeling very odd. It was as if my head were disconnected from my body. My arms & legs would function, propelling me forward, going through the motions, while my mind was running a continuous film of every day, every hour of the life of us as a family."

When he was around other people, he wrote, the film would pause. He could drink beer, and go dancing, and banter with his companions from the hostel. "I would deeply invest myself into the conversations and enjoy the outings that I was invited on, as if nothing had happened. To everyone else, I was an overtly satisfied person, a travel writer using the occasion for an extended vacation, possibly a story."

Longo assumed, he wrote, that no law-enforcement authorities were searching for him in Mexico. He wrote that he didn't even consider the possibility of capture, which is why he remained in Cancún rather than traveling to a more secluded locale, and did not disguise his appearance in any way. The *New York Times* alias was his only safeguard, and that was used less to evade the law, he said, than to avoid speaking about his own life.

On New Year's Eve, four days into his trip, he joined a gang of his hostelmates and danced and drank until three in the morning, after which he swam nude in the ocean with a tall blond woman from Norway. The following day he snuck onto a private white-sand beach behind one of the more exclusive hotels and met a tall blond woman from Sweden. They, too, decided to go for a swim. "I stripped down to my faded navy trunks," he wrote, "while Monica slipped out of everything."

For the second day in a row, though, the relationship progressed no further than skinny-dipping. "I still mentally refused to believe that I was no longer married," Longo wrote. When Monica presented him with an overt offer for romance, Longo told her, "Thanks, but no thanks."

The refusal so startled the men from the hostel who'd joined Longo on the beach that one of them asked him, point-blank, "Mike, are you tutti-frutti?" He replied that he wasn't. He preferred to select his dates, he said, rather than be assaulted by them. He was unsure, though, if the guys believed him.

The attentions of women, Longo wrote, were not at all what he wanted. According to his letter, he'd come to Mexico "to get through a period of grief," quietly and peacefully. Instead, he "witnessed extensive drug and alcohol abuse, was essentially offered sexual exclusivity by four different women w/o prompting, was pressured to buy everything from cocaine to prostitutes, saw enough vomiting to fill a couple of barrels, and encountered the local police in their finest." The police incident occurred when one of his hostel friends was forced to bribe an officer $35 to avoid arrest for urinating on a bush. (Cancún authorities, Longo wryly noted, "evidently have a great amount of sympathy for the local shrubbery.")

He was miserable. Longo devoted paragraph after paragraph to a diatribe against the beach resort's depravity. This section, written in a vigorous, sermonlike style, was the most explicitly emotive of his entire letter. Never for a sentence, though, did he acknowledge that stealing a car and a plane ticket, not to mention possibly murdering all four of his family members, three of them children, was the least bit morally hazy. The bush-peeing incident, he added, was his "last straw"—he needed to escape from his escape.

Some respite came when an older couple, schoolteachers from Britain, took a room at the hostel. They were the first married pair Longo had met since his arrival in Mexico, almost a week earlier, and he quickly befriended them. They played cards and chatted. The couple was on a yearlong tour of North, Central, and South America. They'd sold their home and quit their jobs to make the trip possible. Longo was enamored. "They were fulfilling the kind of life," he wrote, "that MaryJane and I had often talked of doing someday."

He joined the couple for dinner, and the discussion eventually came around to Longo's profession. In his letter, Longo wrote out the entire conversation, at least as he recalled it. After Longo said that he wrote for the *New York Times,* the British man—Longo didn't mention his name—commented that he'd always wanted to be a writer. "How," he asked Longo, "did you decide that journalism was what you wanted to do?"

At this spot in his narrative, Longo broke away from the dinner story to deliver a succinct discourse on his history of lying. "I've always had a bit, well rather a lot, of trouble telling the unadulterated truth," he wrote. "There's been the little white lies, the avoiding-unnecessary-conflict lies, the sparing-other-peoples feelings lies, the wishful thinking lies, the for-your-own-good lies, and now the forthcoming lies that were just lies of fantasy." He insisted, however, that he was unskilled in deceit and found lying to be neither enjoyable nor easy. "It's not second-nature for me to look someone, even a stranger, in the eye & flat out lie to them," he wrote.

Nevertheless, he proceeded to do an excellent job of it. Until his dinner with the British couple, he'd found that being Michael Finkel was relatively unchallenging. To most people he was simply Mike the journalist. Nobody cared to inquire much further; the only question he was repeatedly asked was, "How much do you get paid for an article?" to which he always replied, "I don't talk about money." Now, it appeared, his facade would really be tested.

He answered the man's question about his decision to pursue journalism. "Actually," explained Longo, "through school I never thought of getting into the journalism profession at all. I hated writing. Anything to do with writing instantly became my worst subject. But my teachers saw potential. No matter how much I hated it, something about the words on the pages showed some promise."

"So why," asked the British man, "are you a journalist then?"

"I was in college," Longo replied, "University of Michigan." He said that he was working on a business management degree—in truth, he'd never attended college—when it struck him that he was not the type of person who would ever be satisfied with a desk job. He reevaluated his life's goals and determined that he needed to explore the world. Travel was his true love. So, despite his dislike of writing, he changed his major to journalism. He assumed that he'd one day learn to tolerate and perhaps even love the writing part.

"Did you learn to love it?" asked the man.

"Nope," said Longo. "I hate it, but I get to travel."

Here, once again, Longo inserted a short aside into his letter.

By this stage of the conversation, he wrote, his initial fears—that he'd be unable to answer the questions; that he'd be exposed as a fraud—had dissipated. "For the first time," he wrote, "I was beginning to assume the role, feeling the part." He found it strangely pleasurable.

"So how," came the next question, "did you get to the point you are at currently?"

To answer, Longo whipped up an instant Horatio Alger tale. While in college, he explained, he earned extra money delivering newspapers. He'd be up at three in the morning, distributing the *Wall Street Journal* and the *New York Times*. His work ethic impressed the bigwigs of the home-delivery trade, and upon graduation he was asked to manage the Detroit district of paper carriers—a job he once actually held. Soon, with a promotion, he was in charge of the entire Midwest. In this capacity, he was introduced to executives and editors at both papers.

He began writing articles, on a freelance basis, and sending them to a top editor he knew at the *Times*. "Low & behold," he wrote, "I was now being published in The New York Times & Times Mag. on a consistent basis. The rest, as they say"—and he actually claimed to have said this—"is history."

The British couple apparently believed him, for the conversation continued with a discussion of the various topics he had written about. Longo handled this part—speaking about the specifics of his writing career—with eloquence and ease, he wrote. And then, appropriately enough, for this happens to journalists all the time, the British gentleman mentioned that he and his wife had so many of their own fascinating stories to tell. All he lacked, alas, was the name of someone he could contact.

"I took the hint," Longo wrote, "and offered to possibly provide him w/ a name & perhaps a kind word to my editor." He gave the couple an invented e-mail address, MFinkel@NYTimes.com, and they thanked him profusely.

Several other hostel guests happened by, chairs were pulled up, and the topic of conversation shifted. "I had successfully bluffed my way through my first round of personal questioning as Michael Finkel," exulted Longo. Beers were ordered and consumed, and as the evening rolled on, Longo's mind began to wander.

"I sat there half-heartedly joining, and half daydreaming of what the real life of Michael Finkel must be like," he wrote. "I've learned enough in life to realize that no life or career is as fantastic as you might imagine, but I couldn't help picturing how my life would have been if I had taken whatever steps the real Mr. Finkel took to attain the position that he now held."

FOURTEEN

THE REAL MR. FINKEL absorbed this letter with no small measure of creepy fascination. As I read, I was struck by an odd feeling of detachment, thinking of Longo thinking of me. It was both riveting and uncomfortable; I imagined it might be something like viewing an unauthorized, low-budget movie of your own life.

Longo's impersonation wasn't entirely untrue. I actually did earn an undergraduate business degree—I majored in finance at the University of Pennsylvania—and I did experience a revelation that inspired me to reject my studies and try my luck as a globe-trotting reporter. Longo's dread of being tethered to a desk very much mirrored my own.

Less accurate, I feel duty-bound to admit, are the parallels between Longo's sex appeal and those of his alter ego. I can state with certainty, and some sadness, that any time someone answering to the name Michael Finkel has been skinny-dipping with Scandinavians, I was nowhere around.

Utterly false, and provoking a wince from me each time he mentioned it, was Longo's professed hatred of writing. I don't feel comfortable claiming the opposite—that I simply love to write—though my relationship with the craft is so neurotic and thorny, and has extended over such a significant portion of my life, that love may, indeed, be the best single word to describe it.

I grew up in comfort and stability in Stamford, Connecticut, a fifty-minute train ride from New York City. Both my parents were from the Bronx—hardworking, left-leaning, strict but not unreasonable. My mom taught learning-disabled students in elementary schools; my dad was an executive in the insurance industry. They've now retired to Colorado. My younger sister, Diana, is my only sibling. She works for the Minnesota Department of Corrections, guiding young offenders on meditative trips into the wilderness.

We were a family of readers. It was not unusual for the four of us to retire to the living room after dinner and sit together in silence, everyone with a nose in his or her book. I wrote my own books as well, their covers constructed of cardboard and wrapped in colored fabric. My mother still has two of them. I composed *Avalanche!* while I was still in elementary school, and its title page is indicative of my ambitions: "Written by Michael Finkel. Illustrated by Michael Finkel. Published by Minkel Publishing Company." In a journal I kept at age eleven, I wrote that I'd made a critical decision—I wanted to be a writer when I grew up. (My backup choice was "mad scientist.")

By the time I was in college, though, I had changed my mind. I worked on the university's newspaper, and enjoyed it, but what I really wanted to do was earn money. Hence the business degree. It wasn't until my senior year, in 1990, that I considered pursuing a job in journalism. The catalyst for this was the *New York Times Magazine*. I was enrolled in a writing class, the first I'd ever taken, and was given an assignment to compose a piece that fit the themes of a column that used to run in the *Times Magazine* called "About Men." I wrote of undressing in my high-school locker room, and the mild trauma of exposing my delayed puberty to my peers. My professor liked it, and encouraged me to mail the essay to the *Times*.

A few weeks later, I received a phone call from an editor. He said, to my astonishment, that he wanted to publish my piece in

the magazine, and pay me $1,000 for it. This came as I was apply-
ing for investment banking jobs on Wall Street. The modest wind-
fall didn't immediately change my plans for employment, but it
did fund a postgraduation adventure: I spent the summer bicycling
across the United States.

I went with a friend, and we pedaled nearly five thousand miles.
We crossed ten states, starting in Oregon and finishing in Virginia,
and camped out for seventy-four nights. The trip changed my life.
When I returned home to Connecticut, I wrote an article about it,
which was published in the travel section of the *Times*. I'd also realized, while perched upon my bicycle seat, that I no longer wanted
to be a banker.

My first job in journalism was as a low-ranking editor for *Skiing*
magazine, based in New York City. I appreciated the city, but every
time I traveled to the mountains, I felt at home. In December of
1992, shortly before my twenty-fourth birthday, I resigned from
my staff job at *Skiing,* with the understanding that I could still con-
tribute articles to the magazine, and moved to the mountain-
ringed town of Bozeman, Montana. It's been my home for the last
twelve years.

I haven't moved, but I have traveled. For a while I wrote brief articles for *Sports Illustrated* on unusual competitions—hot-air-balloon
racing; competitive skydiving; the world championships of pinball.
Later, I began writing travel stories for *National Geographic Adventure*—
rafting down a Central African river; skiing the Canadian Rockies;
crossing the Sahara desert.

In March of 2000, on assignment for *Adventure,* I visited Haiti
with a photographer and close friend of mine named Chris Anderson. We were working on a piece about hiking in the Haitian coun-
tryside. While there, Anderson and I spoke with many people who
were so desperate to abandon the poor conditions in Haiti that
they were willing to risk their lives by piling onto rickety boats and
trying to cross hundreds of miles of open water to reach the

Bahamas or the United States. This seemed like a vastly more significant story than an article about hiking. I wanted to document one of these journeys, and I wanted to do so in the most vivid way possible: by actually making the crossing on a refugee boat.

The *New York Times Magazine* agreed to publish the story, if I could pull it off. It would be a smaller piece inside the magazine, not a full-length cover feature, but this was good enough for Anderson and me. We flew back to Haiti.

It took us several weeks to gain the trust of a boat captain, but we eventually managed to secure passage on a twenty-three-foot craft named the *Believe in God*. It was made of scrap wood and powered solely by two small sails. The boat could comfortably fit maybe eight people. Including Anderson and me, forty-six were aboard. A bucket served as the toilet. To prevent the boat from becoming top-heavy, everyone but a few crew members had to serve as human ballast. We spent our time packed into the boat's hold, where the heat was stifling. Within a day, almost everyone was seasick, and several people were so dehydrated they were barely conscious.

In a nod toward safety, I carried with me, hidden in my pack, an emergency radio beacon. Triggering it would send a distress signal, via satellite, to the U.S. Coast Guard. There wasn't nearly enough fresh water on board the boat, and by the second day of the trip I was terrified that we'd all die of thirst. Before I set off the signal, however, a Coast Guard ship spotted us. The *Believe in God* was heading straight for a shallow reef—our crew had neither maps nor navigation equipment—so a Coast Guard dinghy was dispatched to warn us. Officers looked into the hold, and the trip was over. The Haitians were handed over to Bahamian authorities and then flown back to Haiti; Anderson and I returned to the United States.

The *Times Magazine* concluded that this would be a cover story after all. When it was published, on June 18, 2000, I received a good deal of attention and praise, as well as some criticism—for the stupidity of the stunt; for thinking I could imagine how a Haitian migrant really felt. I was also given another assignment by the magazine, to write about a homicide in Kentucky. Then I covered the violence in the Gaza Strip. After that, I became a contract writer for the magazine, and was sent to investigate the illegal market for human organs, and then to write about child slavery on the cocoa plantations of West Africa.

The rest, as they say, is history. I was fired by the *Times,* contacted by the *Oregonian,* and informed about Longo. I was humiliated by what I'd done and bewildered by the identity theft, and from this strange coupling emerged an irrepressible fixation. I was obsessed with learning all I could about Longo. But to begin rebuilding my credibility, I needed to be sure that whatever I wrote regarding Longo was scrupulously accurate.

This created a problem. My chief source of information—Longo himself—had promised me, over the phone and in person, that every word he spoke or wrote to me was the truth. "I'll be completely honest with you," Longo had said, "if you're completely honest with me." I swore the same. Yet soon after we made this pledge, Longo also admitted, in his letter, that he was an habitual liar.

I tried to resolve this contradiction as best I could. To substantiate his Mexico story, for example, I interviewed several people who'd vacationed in the Cancún area with Longo. Tom Taff, a fifty-two-year-old from Minnesota, had stayed at the same lodgings as Longo for four days and spent time with him on a guided tour of Mayan ruins, where they passed an hour together chatting atop a pyramid.

"He seemed intelligent," Taff said. "Nice, clean-cut. He said his name was Michael Finkel and that he worked for the *New York Times.* He said he was writing an article on Mayan mysticism. When he told me his name, I thought, hmm, that's a Jewish name. He didn't look Jewish. But I have seen red hair on Jewish guys. And he seemed like a journalist—he was taking notes, constantly writing. He talked about his other stories. I believed him. He said he was single. He said he had to do too much traveling for his job to have a long-term relationship."

Tom Dunstan, a twenty-three-year-old from Britain, hung out with Longo socially for several days. "He introduced himself as Mike and said he worked for the *New York Times,*" Dunstan told me. "He was having a fine time with us; we were smoking joints and drinking booze, diving off cliffs in the jungle. I have photos of him and me with our arms around each other. He was totally cool—I really enjoyed his company. I wanted to stay in touch with him. We talked a lot of politics. We talked about women quite a lot. He said that he used to have a wife. I said, 'What happened?' He said, 'I got rid of her.' She was cheating on him, he said. You're not going to think, 'I'll bet that guy killed his family.' He was well-spoken, polite, obviously intelligent. It made perfect sense to me that he was working for the *New York Times.* He had a good sense of humor. He'd point out chicks like we all would, but he was a perfect gentleman. He always bought drinks for women. He was very respectful."

So the fundamentals of Longo's story were accurate. He did become Michael Finkel of the *New York Times.* But most of his tale could not be confirmed. In Mexico, he claimed, his internal life was chaotic—his thoughts, he wrote, were constantly grief-filled and frantic. But no one I interviewed said he appeared to be in anything less than the highest of spirits.

Longo himself, in his letter to me, admitted that his specialty was spinning phony tales around genuine details. "For me to be able to tell an untruth," he wrote, "it has to have some basis in reality, something that I have experienced on some level of life, or at least be a topic of some familiarity." That's one reason he may have felt comfortable impersonating me: He was already knowledgeable about my work.

Longo insisted, repeatedly, that his arrest in Mexico had inspired him to change his ways, that he was no longer dishonest. This was another of his unverifiable statements. For now, at the early stages of what I could sense was going to be a protracted relationship, I decided to continue absorbing whatever Longo wished to say, without offering criticism that might scare him off.

Once Longo felt comfortable with me, I assumed I'd see more of the personality he had displayed in Mexico—the quick-witted charisma that had apparently charmed everyone he'd met. This was the nice-guy component of Longo's character, the 92.88 percent that seemed to have mesmerized even his wife. First I would study this part. Then I'd search for the rest.

FIFTEEN

DURING HIS SECOND WEEK in Mexico, Longo finally met a woman he didn't want to reject. Her name was Janina Franke. When she checked into the hostel, Longo couldn't help but notice her: She had fluorescent pink hair, a large tattoo of a feather decorating her left shoulder, a ring piercing her right eyebrow, and (Longo observed when she returned his smile) a shiny silver stud in the center of her tongue. All this, noted Longo, ornamented a "very attractive" body.

But none of that interested him, he wrote. What he liked about Franke, who'd arrived from Germany, was that she hadn't come to Cancún simply to drink beer and loll on the beach. She'd arrived, she told Longo, to explore the nearby Mayan ruins. And—here's what really caught Longo's attention—her plan was to photograph these ruins, in hopes of advancing her fledgling career as a professional photographer.

This was too big of an opening for Longo to resist. He informed Franke that he happened to be a professional writer, one whose travel pieces frequently appeared in the *New York Times*. He mentioned that he, too, had developed an interest in Mayan culture. Franke said that she disliked the feel of Cancún, and planned to travel down the coast to the town of Tulum, where there were fewer tourists and more ruins. Longo replied that he'd also become frustrated with Cancún and was hoping to find a more peaceful spot to explore. Franke said that her dream was to become a traveling photojournalist. Longo said that he'd been struck by a similar dream, and had made it come true. He hinted that he might be able to help her out. The next morning, the two of them were on the bus together, heading toward Tulum.

For a while, during the ride down, Longo was giddy. "I thought that perhaps this is how a life of adventure would be," he wrote. "A journalist on a quest for that untold story." He was impressed by Franke's cameras, two Hasselblads and a Canon, and figured that she really was a photographer. (It had crossed his mind that she might also have been pretending.) He told Franke that he'd once been married but was now divorced; he said he'd never had any children. He spoke with her about a possible collaboration. Franke would take the photos, and he'd come up with the ideal story, something that the publications he wrote for would absolutely love, perhaps a piece that combined Mayan history and adventure travel.

At worst, he promised Franke, they'd sell the piece to the *New York Times*. This was only if, by some fluke, *National Geographic* didn't leap at the opportunity to print it. Franke was elated by her good fortune. She even sent an excited e-mail to her mom about her big break.

Longo paused the flow of his letter at this spot to more fully explain his actions. "I didn't feel as though I was misleading anyone, especially not Janina," he wrote. Instead, he sincerely felt as though he were aiding her career. Without his prodding, he wrote, without the enthusiasm he instilled in her, her dream of photographic success would likely atrophy and die. "I saw myself as being the one to provide her w/ a new close-up lens, to draw her closer to her lifelong goals, giving her purpose and hope."

Longo was not delusional. He never actually believed he wrote for the *Times*. In moments of excitement, though, he did appear to think he could fool not only tourists in Cancún but also magazine editors in the United States. Perhaps, by using Franke's photos and the byline Michael Finkel, he could publish a real article. He would mail all the materials from Mexico, and with a bit of luck, he'd receive a paycheck. This would ease his financial concerns.

"You never know," he wrote. "With even mediocre writing combined w/ excellent photography, we might just be able to pull it off. My teachers always said that I had untapped talent, now was my chance to drill down and see what came out. And besides, at this moment I truly did feel like I may have been leading the life of a true travel writer. I think that my visual perception even began to make a transition from everyday sight to artistic interpretation of everything in view. I no longer saw just a busload of people, I began to peer beyond the faces to look into the history behind the wrinkles, or pained expressions. I didn't just see another lonely guy walking down the highway w/ a guitar. I wanted to know how his life transpired, what promted him to learn to play the instrument, and what motivated him to get up early this morning, pick up his music maker and head out for the long stretch of highway. . . . I couldn't wait to get started. There were creative juices flowing that I wasn't aware were ever there in the first place. I was even anxious to get off the bus to find a store where I could buy a notepad, or paper of any sort to begin my new career."

They left the bus in Tulum, but before rushing off to purchase notebooks, they needed to find a place to stay. January is a popular month for Mexican beach holidays, and all of Tulum seemed booked. After being turned away at several places, they hit upon some luck—one last cabana, a stone's throw from the Caribbean, was available at a low-budget spot called the Santa Fe. They took it. The room was rustic: concrete floor, bamboo walls, palm-thatch roof. No furniture, no toilet, no electricity; nothing at all, in fact, except one not-particularly-large bed.

Franke set her luggage down, looked about the cabana, and

said, matter-of-factly, that the arrangement was fine. Longo nodded his agreement, but inside he was panicked. He'd married young, and was inexperienced when it came to women. "Outside of MJ," he wrote, referring to MaryJane in his usual style, "I had never even slept in the same room w/ a single female, much less the same bed."

He had come to Mexico to grieve, not to flirt—the whole reason he'd teamed up with Franke, he wrote in his letter, was to keep other women away. Now he'd ended up in an uncomfortable situation. But he was able to convince himself that everything was fine. He and Franke were nothing more than business associates, working together on an important assignment. So Longo, too, set down his bags.

They took a walk on the beach, and while they strolled, Longo was plagued by troubling thoughts. "I couldn't help but think how much I longed for MJ to be walking next to me," he wrote. He envisioned holding his baby daughter, Madison, on his shoulders while she pulled at his hair. He could almost see Zachery and Sadie racing ahead of him, scrambling on the dunes. "Instead," he wrote, "I was walking side-by-side w/ a woman that I hardly knew, who had no inkling of the thoughts in my head at this moment."

The walk ended with a dip in the sea. Once in the water, Franke became bouncy and playful; "the spring inside her personality," wrote Longo, "was evidently freed." Longo responded to her overtures, he said, with quiet standoffishness. He told her that he wasn't looking for that kind of relationship. Relax, said Franke. She didn't want a relationship either. All she wanted was a little fun. To demonstrate, she removed the top of her bathing suit. "My barriers," Longo wrote, "began to lower." Then Franke asked if he'd help her apply some sunscreen. Longo acquiesced.

They practically sprinted back to the cabana. "If there was an opportunity to stop," Longo wrote, "I didn't take it." They fell into bed together.

In his letter to me, Longo attempted to explain why his love-making was not immoral. "There was fundamentally no sin against any matrimonial vows," he wrote. He was technically correct. Longo had rented the cabana on January 8, 2002, three weeks after his family had been murdered and one day after they'd been buried. At the funeral, which was held in Ann Arbor, Michigan, where MaryJane had grown up, there were only two caskets. Zachery and Sadie, who'd been sunk with rocks beneath the Lint Slough Bridge, were placed in one; MaryJane and Madison, who'd been stuffed in suitcases, were buried in the other.

For several days, Longo and Franke worked almost nonstop on their magazine article. They hiked jungle paths, explored underwater caves, and even woke before dawn to beat the tourists to the ruins ("so as to have an unviolated scene that would allow for a more clear expression of art"). They saw monkeys and iguanas and, one time, an alligator. They climbed Mayan pyramids. Franke snapped photos; Longo took notes. "We worked as a well-oiled photojournalist team," Longo wrote.

He seemed to enjoy the role. It made him feel important. "I wasn't just another person, paying the entrance fee," he wrote. Instead, he'd arrived "to put into words the things that others experienced unknowingly." He filled dozens of pages in his pocket-sized notebook with what he termed "literary snapshots." In group settings, he was often completely at ease speaking with people about his career at the *Times*. "It seemed natural," he wrote.

There were periods, though, when he found it difficult to be Michael Finkel—when "reality came crashing down," as he put it. The worst moment occurred while sitting on the beach with Franke, staring out at the water, waiting for sunset so the light would be better for her to take photos. "Flashbacks of a horrific scene replayed on the film of my mind," he wrote, though he didn't divulge anything more specific. He ran from Franke and began to weep. He felt an overwhelming sense of guilt; he kept repeating the phrase "I'm sorry" out loud, over and over. His emotions, he wrote, seemed to shift with every incoming wave: "I cried, I was angry, I was resentful, I was hurt, I was lonely but I wanted to be alone." More than anything, however, he was confused. "Do I keep playing this role, do I escape to a new destination?"

He decided to return to Franke and stick with the role—"at least temporarily," he wrote, "until I found some enlightenment." They continued to work on the assignment. He joined Franke on a trip to a monkey preserve, then to another set of Mayan ruins, then to an area of deepwater pools called cenotes. He'd brought a stack of DVDs with him to Mexico, and sold several of them to tourists in order to pay for the excursions.

Longo felt increasingly uncomfortable. He was pretending to lead a life he could never really lead. He was aware that everyone he spoke with was being deceived. He no longer wanted to touch Franke. "The costume," he wrote, "seemed too weighty." He wanted to properly grieve, to begin what he called "the process of repair." But he didn't know how.

On the evening of January 13, after a day of snorkeling in the cenotes, Longo joined a small gathering at a cabana where some young Britons were staying. One of the guys was a music producer, and he was playing a few of the CDs he'd helped create. Candles were lit; beers were drunk. A joint was passed around. After a while, someone noticed a bright light outside, shining through the slits in the cabana's bamboo walls, moving back and forth. It seemed odd.

A moment later, the cabana's door was kicked open and a half-dozen men, guns drawn, rushed inside. They pushed everyone to the floor and snapped on handcuffs. A flashlight swept from face to face, then stopped on Longo's. Two men grabbed Longo, one on each arm, and escorted him from the cabana.

Outside were several more men, also carrying weapons. At first, Longo wrote, he thought it was a drug raid. Then he was brought to the person who'd apparently directed the operation, a stocky, athletic-looking man with a thick mane of black hair and a sort of movie-star suaveness about him. He was unarmed. He held what appeared to be a sheet of paper in his hands—it turned out to be a photograph—and glanced from the paper to Longo several times.

"Are you Christian Michael Longo?" he asked.

"Yes," said Longo, not bothering to attempt a lie.

"I'm Dan Clegg," the man said. "Special agent with the FBI."

SIXTEEN

THERE WAS A TIME, just after I'd handed in my article on the cocoa-plantation worker, during which I convinced myself that what I'd written was true. My story made the point that life in West Africa was exceedingly difficult, but by blurring the distinction between poverty and slavery, as a few humanitarian agencies seemed to have done, the situation was made worse. Using the word "slavery" might gain people's attention, but it could also provoke a boycott of West African cocoa, which would only increase the level of poverty.

This is precisely what I wanted to say. I'd cheated on the quotes, but I had captured the correct story. My article was true in *spirit*—it was a higher truth than that bound by mere facts and figures—and I was able to delude myself that this was all the truth that mattered.

I'd written the story in a highly stylized form, one I'd never before employed. It was composed as though I were channeling the thoughts of a young Malian boy. For example, when my main character left his home village, his journey was expressed this way: "He walked for 12 days. Then he reached a very wide path that looked to be made out of a wonderful kind of rock. He had never seen an asphalt road before. . . . Along the road, at tiny wooden stalls, people were selling things. . . . Youssouf wanted one of everything. But of course he had no money. Or, rather, he had a little. He had been paid two coins by one of the families whose fields he had worked in. It was the first time he had ever been paid, and when those two coins were pressed into his palm, he felt, well, he felt different. Like maybe he wasn't a kid anymore. Like maybe he was an adult."

The article was more than five thousand words long, all of it in this naive, singsongy voice. Somehow, the experimentalism of the story, the fact that I was already twisting so many journalistic conventions, made me feel as if it weren't so terrible to have quilted several interviews together to create the story of a single laborer.

"It is truth," I wrote in my journal, "just filtered through a sort of prism." I knew what I'd done was against the rules, and I hid my actions from my editor at the *Times,* though I believed I could wheedle my way out of it in the unlikely chance I was caught. But then I did something impossible to defend.

One of the boys I interviewed was actually named Youssouf Malé, and that is the name I bestowed on the composite character. The decision to use Malé's name was more or less arbitrary; I think I just liked the way it sounded, and the surname was the most common one I'd encountered. As it turned out, I made a poor choice.

I had traveled in West Africa with my photographer friend Chris Anderson, and he took pictures of many of the eighty or so people I spoke with. The Malian Association of Daloa had arranged the interview with Youssouf Malé; I bought him lunch, and we talked as we ate. Anderson was at the lunch as well, along with my translator. Malé spoke expressively, but at the time I didn't think he'd become a major character in the story, and I mentioned this to Anderson. Also, the mealtime setting wasn't appropriate for a portrait, and the light outside was no good, so Anderson elected not to take a photo of him. I had forgotten this when I selected the name of my character.

Once I'd finished writing the article, the photo department at the magazine naturally wanted a picture of Malé. When Anderson said he couldn't provide one, some suspicions were raised at the *Times*. This was an opportunity for me to acknowledge my sins, but I'd already carried on the lie for more than a week, while my initial draft was being edited and buffed, and I wasn't brave enough to confess now.

Instead, I amplified my deception. I always carry a point-and-shoot camera with me when I travel, and I told the photo department that I could furnish a snapshot of Malé. But in truth, I didn't have a photo of him, either. I mailed a photograph of another boy who was part of my composite character—one named Madou Traoré. It was a brazen act, but that is what I did. The photo department was pleased. The picture I'd sent, they said, might appear on the magazine's cover.

The article was scheduled to be published in mid-September of 2001. Then came the terrorist attacks of September 11, which immediately became the focus of every American newspaper. My story was pushed to the back burner, and seemed as if it would never be printed. I'd begun to fret about the article and the photo, and was greatly relieved not to have to think about them anymore.

Soon after United States forces invaded Afghanistan, I was asked to cover the conflict for the *Times Magazine*. This was the most important assignment I'd ever been offered. I accepted immediately. On my way to Afghanistan, I stopped in New York to visit with the magazine's editors. They informed me that my West Africa article had been revived. It was no longer a cover story, but it was going to be published in a matter of weeks.

Here was another chance to admit to my lies. But I also found out that the *Times'* fact-checkers had finished inspecting my article. I'd known from previous assignments that I probably wouldn't have to show the fact-checkers my notebooks—their practice was to double-check facts using outside sources or by telephoning me and asking me to read back what I'd written in my notes. I didn't think they ever imagined that a reporter would purposefully circumvent the truth.

This loophole allowed me a measure of confidence. In my West Africa story, none of the boys I'd melded into one had access to a telephone, and I had accurately quoted all the people who were reachable. I was never called by the fact-checkers to read from my notes, so I didn't have to invent any further lies. It was a relief, but not a great surprise, to find out that my article was judged to be clean. Despite the problem with the photo, I now assumed I was home free.

I was handed a copy of the article, complete with my mislabeled photo, to review one more time before it was sent to press. I sat in a cubicle in a quietish corner of the magazine's offices, on the eighth floor of the *Times* building, in midtown Manhattan, and as I reread the piece, a terrible feeling came over me. I described it in my journal as "a screaming in my head." My heart raced; my hands shook. I felt the need to explain to someone what I'd done.

But what could I say? I was about to leave for Afghanistan. Adam Moss, the magazine's editor-in-chief, had just taken me out to lunch, during which he expressed such faith in me that he said I didn't need a specific assignment to cover the war—he trusted my instincts, and thought I was capable of finding great stories with only minor supervision from the home office. If I confessed to any deception, I knew that my status at the magazine would be severely diminished, at a minimum. It wasn't worth it. I'd told my lies, and now, I reasoned, I had to live with them.

I remained silent. I went to Afghanistan, where I spent two months and filed two stories, both of which were well received. I'd cut my teeth as a war correspondent. I'd ducked bullets, survived an incident in which a car I was riding in was partially run over by a tank, and written fine—and thoroughly honest—prose. My reputation at the *Times,* it seemed, was solidified. I'd soon be covering all the biggest stories in the world. My articles would be widely read. It was the greatest job I could ever imagine.

*   *   *

When I returned home from Afghanistan, there was an e-mail waiting for me from the relief agency Save the Children Canada. The
agency was upset by my West Africa article. I'd mentioned Save the
Children by name and had suggested that their work in the region—
counseling the victims of slavery—was perhaps addressing the wrong
problem.

"Save the Children Canada has read your article closely," said
the e-mail, which was written by a woman in the advocacy, research,
and policy department, named Anita Sheth. "We have located Youssouf and are slightly confused by what we hear, the timing of his
stay and your visit etc. Please let us know more details as it will help
clarify some issues."

I was alarmed, certainly, but not ready to admit anything. I
replied with my own e-mail. In the most ingratiating tone I could
muster, I said that I had meant no disrespect toward their agency. I
mentioned that I'd seen Save the Children workers in Afghanistan,
and I effusively praised the organization for its humanitarian
efforts. Then I tried to slide my way out of the situation. "Originally the story was going to be longer," I wrote, "but the events of
Sept. 11 changed the focus of the magazine, and so it was cut. As
I'm sure you know, whenever a piece is cut there is always some loss
of nuance."

Save the Children did not fall for my tricks. A few days later, I
received another e-mail from Sheth. This one explained that the
agency had sent some of their staff members into the Malian coun-
tryside and had found Youssouf Malé. They'd interviewed him at
length, and learned that his story and mine did not match. "This
information leaves us wondering about the NY Times article and
the details recorded therein," said the e-mail. "At this point we
would appreciate any information you can provide us with in clearing up this matter."

Now I was more than alarmed. But again I refused to yield. This
time, backed into a corner, I tried to sneak out behind a smoke

screen of verbiage. I wrote a long e-mail in which I attempted to explain my actions without actually saying I'd done anything wrong. "I wanted to do something a little different," I wrote. "I was really hoping to compose a story that would sing, in a way—that would have a single, sustained voice, that would really be a story that would carry the reader through. . . . Obviously, this style demands some educated guesses—of course I can't really write from inside the mind of a young Malian. . . . It was my hope that the reader would understand what I was trying to do, and forgive me in the spots where the story did not work. Basically, I wanted the sum of the story to be greater than its parts."

This only made matters worse. In the two e-mails that followed, Sheth wrote that "Save the Children is shocked by the details revealed to them about the story construction," and then, far worse, she disclosed that the agency had discovered the one piece of truly incontrovertible deception: "From what we have gathered, the picture you have identified as Y. Male is reported to not be him."

I was caught, and Save the Children was threatening to expose me. I was now gripped by full-bore panic. I considered flying to Toronto and attempting to bribe Save the Children—perhaps a sizable donation would convince them to cover up their findings. I remember thinking that $10,000 would be necessary. I could empty my savings account and bring them cash, a stack of crisp hundreds. In my journal, I even began drafting a cover letter to accompany the donation. "You work for a relief agency; you obviously have a big heart," the letter began. "Can't we drop the whole matter?"

Instead, I spoke with Sheth on the phone. I admitted to some of my charade, though I tried to share the blame with the editors and the photo department. After some discussion, Sheth and I reached an agreement. If I wrote a letter of apology to the executive director of Save the Children Canada, in which I stated my regret for any harm my article may have caused the agency, Sheth would require only that the *Times* run a correction about using the wrong photo. Neither of us would mention that I'd created a composite character. This arrangement, I thought, would allow me to save my job.

I phoned my editor at the *Times*, Ilena Silverman, and told her that I'd been contacted by Save the Children and that I'd evidently made a mistake. In the mad rush to finish my West Africa article, I said, I must have accidentally mailed the wrong photo. Silverman took the news well, far better than I'd expected. She said that the magazine would run a correction, but that it wasn't a big deal. She told me not to worry.

But when Silverman informed Adam Moss, he was less understanding. In fact, he was extremely suspicious. How could I possibly spend weeks writing a richly detailed story about one boy, and then send in the wrong photo of him? He telephoned me and expressed his concern. He asked, more than once, "Does Youssouf Malé really exist?" I assured him that he did. But there was a catch in my voice, and Moss must have heard it. He said that the deputy editor of the magazine, Katherine Bouton, would be thoroughly rechecking my story. He told me to express-mail my notebooks to the magazine, immediately.

My notes, I knew, would sink me. I stayed up most of the night, frenzied with worry. I thought, seriously, about burning all my notebooks. Then I thought about faking them—adding bits and pieces that would make my tale true. I even found the shade of blue ink that matched the one I'd used in West Africa. I attempted to forge one page, but it looked so obviously phony, I didn't bother with a second. Even in my panic, I knew that this idea was beyond my limits.

What I did was cash in some frequent-flier miles and book a flight to New York City. I left the next morning at dawn. I called Katherine Bouton on my way to the airport, and on her voice mail I told her that instead of mailing my notebooks I was delivering them in person. I was already in Minneapolis, changing planes, when I received the message she left on my cell phone in reply. She was still under the impression that this whole thing was a misunderstanding, and that once she received my notebooks everything would be clear. "There's no need to come to New York," she said, cheerfully. "I just have a few questions. But if you're already on your way, it'd be lovely to see you."

I arrived at the *Times* building, on Forty-third Street, and Bouton came down to the lobby to meet me. She took one glance at me—my eyes must have looked wild and frightened—and instead of escorting me upstairs, she took me out of the building. We walked to a restaurant around the corner. I had almost no appetite, but I couldn't slake my thirst. Over cup after cup of iced tea, I told her about the West Africa story. I told her exactly what I'd done; I held nothing back. I came clean.

Bouton was kind to me. We both knew I was in big trouble, there was no pretending otherwise, but she managed, somehow, to seem supportive. She told me to wait an hour and then return to the office. She said she'd relate what I'd just told her to Adam Moss. Just take a walk, she said, and then she fixed me with a strange, worried look—a look that said to me, You won't do anything rash now, will you?—and disappeared into the building.

I walked uptown, to Central Park. I felt numb. There were pedestrians; there was traffic. An empty water bottle was pushed by the wind. A man sold poems at Columbus Circle. Squirrels ran in the trees. The iced tea caught up with me, and I urinated under a footbridge.

I shuffled back to the *Times* building, through the revolving door, and into the lobby where the pillars were decorated with headlines from world events. I phoned Bouton, and she came down and brought me through security. We went into the elevators and up to the eighth floor. We didn't speak. We walked past the travel section, and the book review, and into the magazine's offices, with the big blown-up photos on the walls, past the cubicles, crowded and messy, and I finally asked her, "How did Adam take the news?" and all she said to me was, "Not well," and it was like a doctor saying this is really going to hurt, and my chest felt as though the wind had been knocked from it.

The editor-in-chief's office was at the far end of the floor, in the corner, and I walked there staring straight ahead. I didn't want to make eye contact with anybody. Moss was waiting for me. He ushered me inside, along with Bouton, and shut the door.

I had been in his office a few times before. It was, by *Times* standards, a nice one—a giant desk, a decent sofa, a view of lower Manhattan. We had a sort of tradition, Moss and I. Whenever I returned from an assignment, I'd arrange a layover in New York and stop by the office to meet with him. I had come there after spending time on the Haitian refugee boat, after witnessing the uprising in the Gaza Strip, and, just a few weeks earlier, after covering the war in Afghanistan.

I'd sit outside his office, always feeling a touch nervous, like a schoolkid waiting to see the principal. Sometimes Ilena Silverman would wait with me. It never took more than a few minutes before Moss poked his head out the door and waved me in. He would give me a hug and say welcome back. I'd sit on his couch, and Moss would wheel his chair around from behind his desk and sit near me. He'd usually invite a few other editors to join us. He'd ask me what happened on my trip, and I'd tell my story.

The editors would listen, and then, when I was through, they'd comment on what I had said. My article would start to take shape. I'd feel the weight of the task before me—it was always a long, sleep-deprived struggle to complete a piece for the *Times Magazine*—but after the meeting Moss would tell me to take a friend and go eat a great dinner in the city before I left, and put it on my expenses. That was his way of sending me home. Then we'd hug again and I'd saunter across the eighth floor, slowly, flirting with the women in the photo department, chatting with the editors in the cubicles. I'd take the elevator down to the street, walk through Times Square, through the full floodwaters of humanity, everyone seemingly in a mad rush save for me, and I'd feel like a big shot—like I'd conquered New York City.

This time Moss remained behind his desk. The look on his face, as if I were a stranger, was crushing. The meeting didn't last long. Moss simply verified what I'd done. He didn't yell; his voice remained even and quiet. I remember shrinking back into the sofa, squeezing myself between the cushions. "You are young," Moss said to me, trying to be consoling. "You have a long career ahead of you." Then he paused before letting the hammer fall. "But not here," he said.

Those were some of the last words he spoke to me. I escaped from Moss and broke down in Bouton's office, next door. But I felt suffocated in there, so I fast-walked down the hall, past the book review and the travel section, to the elevators. I fled through the revolving door, away from the building. The Editors' Note was published the following week.

SEVENTEEN

THE VERY DAY that Christian Michael Longo was placed on the FBI's Ten Most Wanted Fugitives list—January 11, 2002—a Canadian woman contacted the bureau and said she'd just returned from Cancún and believed she had seen Longo there. This information was immediately passed on to Special Agent Daniel Clegg, the coordinator of the FBI's fugitive program in Mexico City.

Clegg printed a hundred copies of Longo's wanted poster and sent them to police headquarters in Cancún. The posters were hung on phone booths around the region. Two days later, on Sunday, January 13, at about nine-thirty in the morning, the U.S. Embassy received a call, which was patched through to Clegg. It was a Mexican citizen, a freelance tour guide. The guide told Clegg that he'd escorted an American man through the jungle near Tulum the day before. He'd seen the wanted poster, he said, and was certain he'd guided the same man.

Clegg caught the next flight to Cancún. He landed at 5:30 P.M. and was picked up at the airport by Mexican officers. In total, a dozen officers in four cars, all unmarked black sedans, drove the seventy miles to Tulum. The officers were dressed casually, some in shorts, some in slacks. None wore a police uniform. Clegg wore long pants, a gold-colored shirt, and hiking boots. He didn't carry a weapon, but the Mexican officers had either rifles or handguns.

They drove to a place on the beach called the Santa Fe, where the tour guide said he'd picked up Longo. An advance team was sent in. The team shined flashlights into the cabanas—it was easy to see through the bamboo-slat walls—and Longo was swiftly located. "We have him," the captain of the advance team said, over the radio, in Spanish.

Several officers burst through the cabana's door and brought Longo out, handcuffed. Clegg made sure he had the correct man, then informed Longo that he was wanted for murder. Longo did not respond. "He just looked down at the ground," Clegg said.

Clegg asked Longo where his belongings were, and Longo led him to the cabana he was sharing with Janina Franke. "I was completely confused," Franke later said. She had no clue what the raid was about. She asked one of the officers, who told her that she'd find out soon enough. Look on the news, he said. The next day she checked online and learned of Longo's real name and the crimes he was wanted for. "I cried for hours," she said. "If I thought he was in any way suspicious, I never would have spent time with him. His story was perfectly believable. He was polite, charming, friendly. He was good-looking. I could have fallen in love with him if we'd spent more time together."

Longo's possessions were confiscated from their cabana, and he was loaded into the rear seat of one of the sedans, his hands still cuffed behind his back. He was sandwiched between two Mexican officers. A third drove. Clegg sat in the front passenger seat.

"Were you aware that we were closing in on you?" Clegg asked, after they'd started the trip back to Cancún.

"I wasn't even aware that you were looking for me," said Longo.

Clegg told him that he'd made the Ten Most Wanted list, which seemed to astonish Longo. Otherwise, there was no discussion of the murders. They arrived at the Cancún police station a little after 10 P.M. Clegg explained to Longo his options: He could either be extradited or deported, or he could return voluntarily to the United States.

Extradition or deportation, Clegg said, could involve a lengthy stay in a Mexican jail. He knew one person who'd awaited extradition for eleven years. Clegg described the conditions in Mexican jails. He talked about the food and water; he mentioned tuberculosis and dysentery. Longo stated, emphatically, that he did not want to spend time in such a place. So Clegg made plane reservations for the next morning—a Continental flight from Cancún to Houston, leaving at 7:20 A.M.

They remained in the police station all night. Longo was never put in a cell; he sat in a chair, his hands cuffed at first behind his back and then, later, in front. Clegg bought him a hamburger and French fries, and he watched a little television, in Spanish. Longo closed his eyes a few times but did not sleep. He was wearing faded blue jeans, a gray T-shirt, and sneakers with no socks. He was not interrogated by either Clegg or the Mexican police.

Before dawn, Clegg and Longo were driven to the airport. At the airport, Clegg handed Longo a sheet of paper enumerating his legal rights. At the bottom of the document, under the heading WAIVER OF RIGHTS, it read, "At this time, I am willing to answer questions without a lawyer present." Longo signed his name, as did Clegg.

They boarded the plane before any other passengers. They sat in the last row—Longo at the window, Clegg in the aisle, an empty seat between. The row of seats to the front and side of them were left vacant. Once Longo was seated, his cuffs were exchanged for plastic restraints, which were tied to his belt.

For the first fifteen or twenty minutes of the flight, neither man said anything. Drinks were served; Longo ordered a tomato juice. Then came a meal. When they finished eating, Clegg began to talk. "Listen," he said, "I'm going to ask you a few questions about what happened back in Oregon."

Clegg began by explaining that he had transferred many fugitives from Mexico to the United States. He mentioned that he'd read a report on the Longo family crimes. He made eye contact with Longo and said, "You don't look like a monster to me. You're clean-cut. You look like you would be a good father." He continued. "There's always some rationale, some logic, in my experience, to every crime. The report describes someone who committed a horrendous crime. And you don't look like that kind of person."

And then Clegg told Longo his theory. "You were perhaps sexually molesting your children," Clegg said. "Your wife caught you. That led to a fight. Your wife or one of the children got injured, and you decided to kill them." Clegg stopped there. Then he asked, "Why did you kill your wife and three small children?"

Longo, by Clegg's recollection, answered like this: "The scenario that you just told me could not be further from the truth."

"Tell me why, then," said Clegg.

According to Clegg, Longo responded with this: "I sent them to a better place."

This conversation was not tape-recorded. No notes were written; no other witnesses were present. Even though FBI policy states that fugitives must be accompanied by two officers, Clegg had chosen to escort Longo without backup. By the time he flew to the United States, Clegg had been awake for thirty-six consecutive hours. He did not write his report about what took place on the airplane until four days after the flight.

When they landed, Clegg and Longo waited for all the passengers to exit. Then they were transported to the Houston FBI field office. There, in a basement room, Longo was questioned again. This time, Clegg was joined by two officers who'd flown in from Oregon—Sergeant Ralph Turre of the Lincoln County Sheriff's Office and Detective Roy Brown of the Oregon State Police. They spoke for four hours. "The atmosphere was pretty relaxed," Turre later recalled. "It was a good discussion." Again, though, there was no tape recording, and no notes taken. When Longo was asked, in Houston, "Did you kill your family?" his answer, according to Turre, was, "I don't want to talk about that right now."

"Longo does not admit outright that he had killed his wife and children," Turre wrote in his report about the Houston interview.

"Mr. Longo openly admitted that he killed his wife and three small children," Clegg wrote in his report about the airplane interview.

"There was never any confession or even discussion of that night," Longo wrote in a letter, referring to both interviews.

Longo stayed overnight in Houston, at the Harris County Jail. He was kept on suicide watch, and once again he did not sleep. In the morning he flew to Oregon, escorted by Turre and Brown. On the plane, they chatted about photography and scuba diving.

They landed in Portland about 8 P.M. on Tuesday, January 15. An unmarked police car was waiting for them on the tarmac, and they were driven two and a half hours to Newport, to the Lincoln County Jail. Late that evening, shortly before midnight, Longo joined Turre and Brown in a conference room at the jail and submitted to his third interview. They sat at a round table. Longo was wearing a navy blue jail uniform; Turre and Brown wore suits but had to take off their ties, which are considered risks to jailhouse security. Neither officer was armed, and Longo was not cuffed. There was a pitcher of water on the table, but no one drank from it. This interview, like the one in Houston, lasted four hours, including a bathroom break. Unlike the others, this one was tape-recorded. The tapes were entered into evidence and later made public.

For more than an hour, Turre and Brown questioned Longo about his actions after the crimes—his drive to San Francisco, his escape to Mexico, his impersonation of a *New York Times* reporter, his liaison with Janina Franke, and finally his capture, which Longo described as "a big weight off my shoulders." Turre spoke softly, constantly inquiring about Longo's feelings. Brown was the bad cop, interested only in the facts. Together, their goal was to get Longo to confess to the murders, on tape.

When it came time to talk about the crimes themselves, Turre took control of the conversation. He was aware of Clegg's contention that Longo had sent his family to "a better place" and thought, therefore, that the murders might have religious undertones. His feeling was that Longo may have killed his family out of some desperately misplaced love.

"Would you," asked Turre, "ever allow anyone to intentionally hurt your family?"

"No," said Longo. "My family is everything."

"And if you knew that your family had been hurt, intentionally, by someone else," continued Turre, "would you report that to the police and expect them to do something about it?"

"Most definitely," said Longo.

"Okay," said Turre. "Would you ever ask someone else to hurt a family member for you?"

"Never," said Longo.

"Okay," said Turre, using this word, a favorite of his, as a sort of verbal balm. He was preparing his trap, and wanted to be as gentle as possible. "You know I've looked at all your family. Okay. And I know that the manner they left this world in was not a brutal manner. Okay. It wasn't a bloody scene, it wasn't a brutal act. I think it was the act of a desperate man who didn't know which way to turn, and thought that he was doing the best he could."

Turre then spoke for a minute about what he called "classic familicide"—a form of murder in which a father kills his family members not out of hatred, but because he feels unable to adequately care for them and doesn't want to see them suffer. Turre said that everything he'd heard from Longo seemed to support this idea.

"I believe that you truly loved your family," Turre continued. "I believe that you truly still love your family. And I believe that the only reason that you took their lives is because you felt like the life that you were providing them here on Earth was not what they deserved. And that you knew in the next life they would be in a better place.

Now in a lot of respects, that is something that a loving father would want them to see—a better life than the one that he is providing them with. The reason I asked you, 'Would you ever allow someone else to intentionally hurt your family,' is I know you wouldn't. Because I believe you truly loved your family. Am I wrong in my assessment? I told you what I think happened. Am I wrong?"

On the tape, Longo's sobs—gasping, sniffling sobs—are all that can be heard. Turre allowed them to continue for some time. Then he spoke again, not much louder than a whisper. "Or am I close?" Turre said. "Or am I right on?"

Longo continued to weep. Then, after a while, he composed himself. "I don't know," he said, "if I can safely comment on that right now."

Turre pushed harder. He shed a little of his tenderness. "I mean," he said, "you can withhold the, 'Yes, I did it,' feeling that there is safety in withholding that. But Chris, I think it's beyond that now. It was beyond it before we even went down to talk to you. Because there is no one else on this earth who would have done that to them. Okay. And left them where you left them. The main thing we needed to know, when we talked to you, was why. And how. It's not so much, 'Yeah, I did it.' I know you did it. Roy [Brown] knows you did it. You know you did it."

But of course, both officers understood that without a clear confession, they could not be certain Longo did it. So Turre kept going. "You can withhold how you did it," he said. "The fact remains you did it. And we talked about ownership. Taking responsibility. And Chris, there eventually has to come a time that you do that. I think now is the time. And it is only two little, three little words. They can be quickly spoken."

Longo, still sobbing, did not take the bait. "I'm going to wait," he said. "I'm sorry."

Turre wasn't ready to give up. He tried a slightly different approach. "Okay," Turre said. "Let me ask you this. Could anyone else have killed your family?"

Longo paused. He sniffled. "That's a loaded question," he said. "I'm going to wait."

With this, the officers surrendered. Longo was obviously not going to be cajoled into confessing, and so the interview was concluded. It was the last time Longo spoke with police investigators. A week later, on January 23, 2002—Longo's twenty-eighth birthday—the Lincoln County district attorney, Bernice Barnett, announced that, due to the heinousness of the crimes Christian Longo had been accused of, she would be seeking the death penalty.

PART THREE

LOVE

EIGHTEEN

THERE WAS ONE thing I wanted to get straight between Longo and me. We both knew, from the first minutes of our first phone talk, that we were spiraling around the central topic, and that it was only a matter of time before I'd have to ask him about the murders. I forced myself to remain patient during our initial phone call, and then, when I traveled to Oregon and saw him in the Lincoln County Jail, I held off through the majority of our visit. But as I sat at the booth, studying his face, the urge to broach the subject itched at me with every conversational pause.

Finally, as the visit drew to a close—this was just before Longo displayed the letter and said he'd decide whether to mail it to me—I gathered my nerve. I looked him squarely in the eyes. I spoke clearly and assertively. "Chris," I said, "did you do what you are accused of doing?"

His face remained composed. It was as though he'd been waiting for me to ask this. He was silent for a moment, and I felt he was selecting his words carefully. "I can't answer that right now," he said. "But I think you know." And then he winked at me, winked his left eye, slowly and obviously, as if to say, Hey, our conversation might be monitored so I can't say anything directly, but there's your answer.

I thought it was an effective one. He'd avoided incriminating himself, and at the same time, he hadn't lied. He easily could have said, "No, of course not," but instead he said, "I think you know." And by this point, I did know. I'd read every word about Longo that had been made public—police reports, search warrants, media dispatches, court rulings. I'd spoken with Longo himself. I knew. He was guilty. The evidence against him was overwhelming.

As I sat there, on the visitor's side of the glass, with the afterimage of the wink sharp in my mind, I felt tugged in opposing directions. Here, in front of me, was a person who deserved nothing but contempt: a man, apparently sane, who'd murdered his own family. And here, as well, was a perceptive prisoner who seemed willing to explore the roots of his crime—and, quite possibly, help me restart my life. Part of me wanted to run, but more of me wanted to stay.

In his letter about his experiences in Mexico, Longo had mentioned the crime a few times, but on each occasion, his writing abruptly shifted from an emotional, first-person account to an odd, detached third-person voice. He wrote of "the terribly unnecessary demise of the lives of a wife and three beautiful children" and noted that "a much loved family was suddenly no more." He talked about "the disaster" and "this catastrophe" and "that night." He mentioned "a tragedy that has recently taken place." But never once did he use the word "murder" or "killing" or "homicide."

Then, in June of 2002, a month after he'd mailed me his Mexico tale, Longo sent me another letter. This one was also exceptionally long—fifty-seven pages. He gave it a title: "Wrong Turns." The letter contained an intricate, at times obsessive, accounting of every mistake Longo felt he had made in his youth, starting with an incident in ninth grade in which he stole a roll of quarters from his dad's dresser.

He wrote about paying for a PG movie but sneaking into an R (*Tango & Cash*); removing a jarful of vodka from his parents' liquor cabinet; getting into a brief fistfight in the high-school cafeteria; using his dad's credit card to order a bouquet of roses for a girl; and, after receiving a D in biology, leaving a message on his home answering machine in which he pretended to be the biology teacher calling to correct a mistaken grade—an early impersonation that failed entirely.

This was, of course, Longo's personal selection of his misdeeds. Throughout our correspondence, I attempted to verify as much of what he told me as possible. As with the letter describing his time in Mexico, virtually everything that could be checked turned out to be accurate, though this still left a large amount of unconfirmed information. I also didn't know which events Longo had omitted from his life story. If he wasn't lying, it was likely, I realized, that he was at least skewing his narration to showcase himself in the most sympathetic possible way.

On the final page of "Wrong Turns," almost as an afterthought, Longo brought up the death of his family. Here, for the first time, he set aside the passive syntax and issued a direct statement. "I didn't commit the act," he wrote. He did, however, feel guilty—"this whole incident is my fault"—but only "for not being home to ultimately protect."

By the time this letter arrived, Longo and I had established a regular weekly telephone conversation. On Wednesday evenings, most of the inmates in Longo's section of the jail attended church services. Longo's segregation status prevented him from joining, but this was an ideal time for him to use the phone. I always made sure to be home, awaiting his call, with a fresh cassette tape in my telephone recording device.

Longo's declaration of innocence necessitated further discussion, but I had to be careful. Neither of us knew who might be eavesdropping on our talks, and Longo tended toward circumlocution when anything sensitive was brought up on the telephone. So when inquiring about his "not being home" statement, I began generally.

"In all your letters," I asked, "have you been honest with me the whole time?"

"I have been painfully so," he said. "More so than I probably should have."

"Everything you've written to me is true?" I asked again, just to make sure.

"It's all one hundred percent factual," Longo said.

"There's nothing that you want to take back?" I prodded.

"No," he said. "I've been honest about everything."

Before he'd mailed me "Wrong Turns," I had told Longo that I felt comfortable maintaining the legal assumption of innocence. "I'm keeping an open mind," is how I phrased it. This was easy to do so long as the murders were not discussed. Now, I told him, I felt as though I were "doing some crazy yoga move, bending over backwards to believe you."

"I understand that, and I hate to have you do that," Longo replied. He informed me that for the time being, with his trial still ahead, there was nothing further he could say about the matter.

"Assuming that you're telling me the truth," I added, attempting in as kind a way as possible to imply that an innocent man whose family had just been murdered would not likely flee to Mexico instead of calling the police, "you did some really stupid things."

"Yup," is all Longo said. A minute later we were disconnected.

The conversation must have struck him, for he wrote me a brief letter, only seven pages, a few days after this talk. "You had inquired over the phone about my being honest in everything thusfar," he wrote. "If you have specific concerns, as you seemed to imply, please forward those to me. I don't want there to be any cause for misunderstanding, much less suspicion of dishonesty, between us."

In all his letters, this was Longo's one unvarying theme—the need for complete and unambiguous truthfulness on both our parts. He repeated this so often it became a sort of mantra. The fact that Longo himself was a skilled liar seemed to engender in him an ancillary condition in which he was distrustful of everyone else's honesty.

"I hear stories about journalists taking people under their wings," Longo said during one of our talks. It's a common tactic, he pointed out, for a reporter to insincerely befriend the people he's interviewing, only to "thrash them in a story, which was his whole point to start with."

And yet, just after he shared this opinion, Longo admitted that he felt an immense need to speak with a trustworthy journalist. Nothing that was written about him in the press, he said, reflected his side of the story. When a local newspaper, *Willamette Week,* ran a two-part article on him, it was headlined "The Making of a Murderer." Though his trial was still many months away, the paper didn't bother to add an "accused" or an "alleged" to the title. A photo accompanying the story had been digitally manipulated so that Longo's head appeared warped and twisted.

"I just read something that said I was a monster," Longo lamented to me on the phone. "People think I'm inhuman," he said another time. He referred to what was happening to him as his "monsterification," and he needed someone to help counteract this process. "I feel like I can't be normalized," he said, "until people understand a little bit about who I am."

This was where I fit in. Longo knew that my firing allowed me to dedicate virtually unlimited time and energy to his story. He realized, too, that I'd also just experienced what it was like to be steamrolled by the press, to be branded a liar, and to have your credibility shot. I was in a perfect position, he implied, to listen to him without leaping to conclusions, to pay attention to facts rather than yielding to assumptions.

Longo had asked me in his letter if I had a "suspicion of dishonesty" about his insistence that he was innocent. Well, I did have such a suspicion. So I promptly wrote back. I even offered him an easy way out of his "not being home" proclamation.

"As I told you on the phone," I wrote, "I have decided to believe everything you've told me until or unless proven not true. This includes the statements you made in your final pages of 'Wrong Turns.' As you are well aware, this requires quite a large leap of faith on my part, and I have decided, with no hesitations, to take that leap. I just want you to tell me, once more, in writing, if there is anything you want to take back, or ammend, or tell me that you were speaking figuratively instead of literally (e.g. 'not at home' can also mean not in your right mind)."

Longo's response arrived within a week. "There is nothing that I wish to retract," he wrote. "If something sounds confusing or contradictory I would rely on you to mention it, for clarification. Regarding 'not at home,' that was literal. (Please be careful what you send, it is read by guards)."

Over the next few calls and letters, Longo expounded further. There were many things I did not understand, he said. If I only knew about the pressure he'd been under, and the bad luck he'd endured, and the sacrifices he'd made to provide for his family, then I would realize that harming his family was something he "could not even conceivably do." He said he'd winked at me during the visit because he thought I knew that he was *not* guilty.

He insisted that if I were patient it would only be a matter of time before his innocence was obvious. In the coverage of his case, Longo wrote, the press "has chosen to publish statements that are in no way based in reality." He told me that his actions—fleeing to Mexico; telling people his wife had left him—were all explainable. "There is much more to this case," he wrote, "than meets the eye."

But the way almost everyone else saw it, there was no chance that Longo could be anything other than guilty. When Longo's defense team hired a polling service to gauge local opinions about the murders, not a single person out of the four hundred who were interviewed said that Longo was "definitely not" or even "probably not" guilty. Longo told me that no relative or former friend had sent him a letter of support. Even his parents, in a letter to the district attorney's office, wrote that they realized "the one who may be responsible for murdering half of our family is our own son."

Longo begged me to ignore all this and listen to him. "I always wondered how people could be convicted of a crime & put on death row, despite being innocent," he wrote. He said that if I only knew "the whole, true story," then I'd clearly see that he was not the person who'd killed his family.

Okay, I said to him. Tell me the true story.

NINETEEN

IN THE FALL of 1990, when Christian Longo was sixteen years old and his brother, Dustin, was fifteen, their parents, Joe and Joy Longo, took a one-week vacation to Arizona. Though the boys would be staying with friends, Joe and Joy felt they needed a house sitter, primarily to care for the family dog. They asked around—the Longos had recently moved to Ypsilanti, Michigan, from Louisville, Kentucky—and hired a young woman named MaryJane Baker.

While his parents were away, Chris bicycled home several times. First it was to play with the dog, but then, soon enough, it was to spend time with the house sitter. Whenever he saw Baker, he felt this woozy, head-to-toe prickle he described as "teen-boy-in-love energy." Baker was twenty-three years old. She had curly brown hair, glacier-blue eyes, and, when she released it, a smile that seemed both girlish and profound, and hinted at something mysterious within. She was slender and petite. Her skin had a natural glow; she almost never wore makeup.

Every part of her that Longo saw turned him on—her lips, her teeth, her ankles. ("Her ankles," he wrote, "were perfect.") Even the trace of shyness Longo sensed in her was alluring. During one visit home, he helped her wash dishes. He was overjoyed simply to stand beside her at the kitchen sink, drinking in the scent of her perfume, feeling her hair brush against his arm when she passed him a plate.

He realized, though, that this was only a fantasy. Baker was seven years older than him, and in a serious relationship. "She was unattainably beautiful," he wrote.

Baker had endured a somewhat difficult youth. Her father had left the family when MaryJane, the third of what would become six siblings, was in elementary school. For a while, the Bakers were dependent on public assistance. Two of her sisters moved out when they were in their mid-teens, leaving MaryJane behind to help care for her half sister, Karyn, the youngest in the family. She'd had no opportunity to attend college. At the time MaryJane was house-sitting for the Longos, she was employed as an assistant in a pediatric office, lived at home, and provided financial assistance to her mother.

She found solace, according to Chris, and a sense of family in religion: MaryJane was a devout and enthusiastic Jehovah's Witness. Before she was able to afford a car, she often walked several miles to attend meetings at her congregation's Kingdom Hall. (Her mom had once been a Witness, but was expelled from the organization for what the church deemed moral lapses, including an out-of-wedlock affair.) MaryJane's work schedule at the pediatric office allowed for one day off during the week, and she spent this day driving the neighborhoods of Ypsilanti, knocking on doors and attempting to share the Witness doctrine—that Armageddon will soon arrive, and only the righteous will live forever in the ensuing kingdom of God.

The Longos were also Jehovah's Witnesses. Neither Joe nor Joy was born into the faith—both were raised Catholic. Both grew up in Iowa, too, though Joe Longo is not the birth father of either Chris or Dustin. Their biological father is a man named Steven Steward, whom Joy married after she became pregnant during her senior year of high school. The marriage was not a good one. Steward was a heavy drinker and, according to Joy, physically abusive. Still, out of a sense of duty, she remained married to him, even after the first pregnancy ended in miscarriage. She became pregnant again, but

Steward, who was driving drunk, according to Joy, rolled a car with the two of them in it, and her second pregnancy also ended prematurely.

Finally, on January 23, 1974, in Burlington, Iowa, Joy gave birth to a child, a son she named Christian Michael. The delivery was troublesome; forceps were needed, and the infant was born with his head cut and bleeding. Fifteen months later, Dustin Anthony was born. But, said Joy, the violence from Steward did not stop. Joy recalled that one time, when Chris was three years old and making a mess of his food, Steward hit him in the face, blackening an eye.

This was Joy's limit. She separated from Steward, filed a restraining order against him, and was later divorced. Chris never saw Steward again, and said he has no recollections at all of his birth father and did not have any desire to look for him. (Steward joined the army and later returned to Iowa; he now works as an electrician, has been married to his second wife for twenty-five years, and has had no trouble with the law. He was unaware of the Longo murders—he didn't even know Chris's last name—until contacted by the media. He claimed that Joy's recollection of their marriage was embellished, and insisted that he never once struck Joy, Dustin, or Chris.)

Joy moved with her two infant sons to Des Moines, where her parents lived. She took a job at the customer service desk of a Target department store and was introduced to an assistant manager named Joe Longo. Joe was gregarious and popular; he'd been a high-school homecoming king and a star football player, a wide receiver, at Morningside College in Sioux City, Iowa. He had a gentle demeanor and a reputation for honesty. He didn't smoke or swear, and hardly ever drank. They went to a company Christmas party, danced, and fell in love.

At their wedding, the four of them—Joe, Joy, Chris, and Dustin—all walked down the aisle together. Soon after, Joe became the boys' legal guardian. Chris loved him. The earliest memory of his life is of Joe dressed as Santa Claus, delivering gifts to his new family. "My dad was my idol and hero," he wrote. They had football catches; they played basketball. The whole family took trips to Florida, New York, Toronto, and St. Louis. They once drove across America. Chris can't recall his father raising his voice, not ever. There was no spanking or hitting. Joe and Joy were so intent on rearing the boys in a nonviolent setting that they did not allow squirt guns in the house. "I couldn't ask for a more loving family," Longo wrote.

In 1980, Joe was promoted by Target to a store-manager position, which took the Longos to Indianapolis. They bought a ranch-style house in the suburbs, with an apple tree in the yard, a basketball court in the driveway, and the boys' elementary school next door. Joy worked as a housecleaner for a while, then became a full-time mom. She was contacted by the Witnesses in the usual style—a knock on the door—and found herself intrigued by their beliefs. She began attending meetings at the local Kingdom Hall, and often brought her sons with her. Eventually she decided to join. Chris, who was ten at the time, and Dustin, a year younger, joined with her.

Joe Longo did not approve. His father was a deacon in the Catholic church; Joe himself had been an altar boy. But he didn't want to cause a rift in his family. He noted the positive influences the Witnesses had on his wife—as soon as she joined, Joy ended a two-pack-a-day smoking habit—and so, cautiously and gradually, he started reading the organization's materials. Three years after his wife's conversion, Joe also became a Witness. He ultimately became so involved in the church that he ended a twenty-year career at Target to devote himself more fully to spiritual goals.

Jehovah's Witnesses are Bible literalists. The scriptures, they believe, were channeled directly from God, and Witnesses do not observe any custom not specifically mentioned in the Bible. This includes celebrations of Christmas and Easter. They sometimes interpret 1 Corinthians 15:33—"Bad associations spoil useful habits" is the phrasing found in the Bible translation preferred by Witnesses—as a directive to minimize social contact with those who believe differently (so-called worldly people). Thus, a community of Witnesses can become extremely insular. Not long after the Longos joined, Chris wrote, "all of our friends were Witnesses." He said he didn't miss Christmas because his mom promised to give him gifts throughout the year, not just on one particular day.

By the time he was a teenager, though, some of the strictures had started to wear on him. Longo was a natural athlete, and school coaches in basketball, football, and track wanted to recruit him. But Witnesses often frown on team sports—competitiveness, they believe, can be spiritually unhealthy, as can extended associations with worldly people. So Longo was forbidden to participate. He couldn't even join the Boy Scouts. "On the scale of strictness," Longo wrote, "my parents were at the top." He once tried to sneak a Guns n' Roses cassette into the house, but his mom found it in his sock drawer and threw it away.

The prohibitions on school activities made Longo feel unpopular. He was teased, he said, for being a "goody-goody" and a "square." He was unquestionably bright—his full-scale IQ would later be measured at 130, which is in the "very superior" range, above the ninety-eighth percentile—but his grades were poor. "I lost all motivation towards school in general & quickly grew to hate it," he wrote. In the first semester of ninth grade at North Central High in Indianapolis, Longo received no As. His only Bs were in nonacademic subjects: woodworking, phys ed, and band (Chris played the alto sax). He was given a C in English and Ds in algebra and biology. The biology mark was the one Longo tried to improve by leaving a fake message on his family's answering machine.

Those were some of the last grades Longo earned in a classroom setting. In 1989, when Chris was still in ninth grade, Joe accepted another promotion from Target, and the family moved to a small town outside Louisville, Kentucky. Shortly after, Chris was removed from school by his parents. (Dustin stayed in. "I always thought that he was perceived as the good son," Longo wrote.) Chris continued his education through correspondence courses and home-schooling. After the family moved again, to Ypsilanti, he received his high-school diploma. He did not go on to college; Jehovah's Witnesses tend to discourage the pursuit of higher education.

Joy Longo said that Chris, as a young man, never displayed any predisposition toward violence; he didn't even fight with Dustin. His main problem was a tendency to lie. "We tried very hard to get that out of his character," she said. The incident in which he pretended to be his biology teacher greatly disturbed her. "He wouldn't admit it," she said, "and we knew that it was him."

Girls were another issue. Longo, it seems, was always able to attract them. "I got my first kisses in grade school," Longo wrote, "my first 'tongue' & 'feel' in seventh grade & my first real make-outs w/ petting in my freshman year."

It was for this freshman-year girlfriend, Georgina, that Longo stole his father's credit card in order to purchase flowers. When he was caught for this, his parents were upset about the sexual indiscretion as well as the thievery. Witness youths are not permitted to touch one another, not even to hug. Joe and Joy, quoting from 1 Corinthians 7:36, explained that Chris would not be able to date until he was "past the bloom of youth" and ready to marry. "That was my last girlfriend," wrote Longo, "until MJ."

TWENTY

SOON AFTER LONGO began unfolding his life story, I mailed him a letter, dated July 12, 2002, that included six pages of detailed questions. I was still bothered by his "not at home" claim, and felt the need to test him further on it. If he wasn't home, I wondered, then where was he? Did he have proof? If he didn't commit the crime, had he any idea who did?

He wrote back, though the reply was terse by Longo's standards, just three and a half pages, and the tone distinctly cooler. "I can assume that it's the reporter in you," he wrote, responding to my avalanche of questions, "but relaying that much info to anyone, at this point, would not be an exercise in wisdom."

Longo had apparently arrived at a realization: Though we'd become acquainted through extraordinary circumstances, I was, in the end, just another member of the press, greedy for the salacious details of his life. "Despite your appealing nature, I'm forced to keep your profession at the forefront of my mind," he wrote. "I do realize that you are a journalist first."

Therefore, Longo concluded, I should be treated like any other journalist—if he was going to speak with me, I should pay him for the honor. He hadn't actually talked with anyone else, but other media outlets, he claimed, had offered him "an amazing amount of money" for an interview, in some cases more than ten thousand dollars. (Precisely who made these offers, he didn't say.)

He was aware, by now, that I'd begun to see his story as something more substantial than a magazine article, and was thinking of trying to write it as a book. Longo insisted that, if and when I received money for the book, I hand over a cut to him. He was not interested in keeping any funds for himself. Rather, he wrote, he wished to donate all profits to his parents and a few friends with whom he had debts.

As for my pledge to listen to Longo's story with an open mind—that, it appeared, was no longer valid. "A verbal promise between an accomplished journalist & the closed-lipped subject of a sought-after story is borderline ludicrousness," he wrote. Longo added that he wouldn't continue our discourse unless we'd worked out a financial arrangement. Until then, he wrote, "I'm forced to clam up."

Paying Longo for the story was out of the question. The moment money changed hands, my work would be compromised; it would mean, in effect, that Longo and I were partners. When speaking with him over the phone, I sometimes referred to the piece of writing I envisioned as "our story," or as "the Chris and Mike Project," but it was in fact *my* project, and Longo knew it. He would have no authority over the prose, no opportunity to view anything in advance, no chance to make editorial alterations. On the other hand, he was aware that if he halted communication with me, there would be no Chris and Mike Project.

I did remunerate Longo for the postage he needed to mail me letters (some contained more than a hundred pages of material), as well as for envelopes, pencils, paper, and a few snacks from the jailhouse commissary. In total, over the course of our communication, I deposited $180 into his jail account. He never asked me to do this, but I knew he had almost no money of his own. Later, I funded a subscription to the *New Yorker* magazine.

Longo had determined, however, that he should receive a share of the project's earnings, and he seemed fixed on this idea. I decided, at this point, to take a chance and phone one of Longo's lawyers. He had two: His lead counsel was Kenneth Hadley, sixty-four, a local Newport attorney; his co-counsel was Steven Krasik, fifty-six, whose office was a hundred miles from Newport, in Oregon's capital city, Salem. Longo had told me that both lawyers were aware we'd been in contact. Neither man, Longo said, was pleased with our association, but to his surprise they hadn't asked him to halt it, either. I grasped at this small opening and gambled that speaking with a member of the defense team, and formally introducing myself, might somehow help resuscitate my relationship with Longo.

This fantasy was swiftly quashed. I called Krasik first, and he told me that he "hated" my communication with Longo—that's the word he used—and said that neither he nor Hadley would do anything to assist me. He did, however, explain why he hadn't advised his client to break off contact with me.

Krasik said that he'd spent a lot of time with people who were facing either a lifetime in jail or a death sentence. He knew that such a prospect was terrifying. The feeling that can overcome an inmate, he said, is that of tumbling down a bottomless ravine. In a situation like this, a connection with the outside world—a way to divert one's thoughts from his imprisonment—is crucial for maintaining sanity. Longo was virtually alone, Krasik pointed out, abandoned by everyone who knew him. Even his parents hadn't visited. They'd just written a couple of letters.

"Except for his attorneys, who aren't in the comfort business," Krasik said, "there's only you." The circumstances that brought Longo and me together were so implausible, it was difficult for Krasik to dismiss them as mere chance. "The two of you were linked before you even knew each other," he said. "And now, from everything I understand, you're his connection. You're his lifeline."

*   *   *

The next letter I wrote to Longo was scrubbed almost completely of inquisitiveness. As was the following one, and the one after that. Instead, I devoted paragraphs to describing my home, including my twelve-acre hayfield—"I have this sort of Jewish-boy-from-the-East-Coast idea that I'll put a few cows on it"—and my chicken coop. "I'd mail you some eggs one day," I wrote, "but I think the result (mailing you eggs) has the potential to be messy. Perhaps hard-boiled eggs." I shared stories of my travels, and explained how I acquired my favorite souvenirs. "I bought my first carpet in Iran; I smuggled it out of the country, illegally, in my ski bag."

I filled him in on the minutiae of my existence: ski trips, hockey games, ideas that came to mind as I jogged. I wrote about the weather. I asked him to vote on a name for my new cat. (He chose Otto.) I shared anecdotes about my Grandpa Manny; I taught him some Yiddish: schmutz, schmuck, schmatte, putz, mensch, meshuga.

My letters grew longer. For months, almost all of the writing I did was for Longo. After the *Times* had finished investigating my articles, several magazines, including *National Geographic Adventure*, were again willing to publish my work. But I had no desire to take any assignments. Instead, I lived off my savings and devoted myself full-time to the Longo project. Sometimes I'd spend an entire day doing little else but writing to him. I virtually stopped keeping a journal; my letters to Longo, each of which I photocopied, essentially became my journal.

I always composed the letters in a single draft, in pen, without any particular objective except to allow Longo into my life—and, I hoped, to inspire him to reciprocate in kind. "I'm just spitting out what's in my head," I explained, "[with] no attempt to strain anything out."

I included in the letters copies of my articles, all the ones I didn't think he'd previously read. I sent him crossword puzzles and initiated a game of chess, one move per letter, with schoolyard-style taunts accompanying each move. I shared the details of some of my dreams—"often I now dream at night about sleeping in trees"—and occasionally made light of Longo's confinement. "I'm not lucky like you," I wrote after pausing my letter to cook breakfast; "nobody brings me food." I mailed what I called "cell-warming gifts"—photos to hang on his walls from my trips to Niger, Mali, Afghanistan, Panama, China, Thailand, and Japan.

Not all I wrote had a jovial tone. I described to him the body-heavy weariness that overcame me when I dwelled on my firing. There were moments, I admitted, when I so fiercely missed the adrenal buzz of a big-story pursuit that the Montana town I lived in, which I'd always adored, had begun to feel like an outpost I'd been banished to.

I also told Longo about my girlfriend. My convalescence had given me the chance to pursue a genuine romance for the first time in years, and I started dating a professor in the math department of Montana State University, named Jill Barker. We'd first attempted to date during my *Times* days; the connection between us, I'd found, had been dauntingly powerful, but the liaison had ended up, as did all my affairs of that period—my attention span erratic, my fidelity intermittent—as an utter disaster.

Shortly after my firing, though, Jill offered me another chance. I kept Longo fully informed, sparing him none of my struggles. "Jill and I have had a few arguments lately," I wrote. "What about? Gosh, mostly about how committed I am to the relationship. Chris, I have to tell you—I'm really somewhat of a failure when it comes to women. I'm 33 years old; I'd really like to have a family one day—but I can scarcely hold down a girlfriend."

As I was writing these letters, I knew that confiding the details of my love life to a man awaiting trial for murdering his wife and children was probably inappropriate. But my letters to him seemed to have a life of their own, one resistant to self-editing. Writing to Longo had become strangely freeing. Whatever I told him seemed safe; even if my letters were being scanned by jailhouse officers, his being locked away gave me the sensation of depositing my words into a vault. And by writing to Longo about my struggles with women, and egotism, and honesty—by digging at my issues until I'd grasped them ably enough to put into words—I at least felt like I was learning something about myself. So I refused to hold much of anything back. If it was on my mind, I usually put it in a letter.

Longo responded to my efforts. He did not follow through on his threat of silence—he just halted his life story. He referred to me as "the ultimate pen pal," and then, as if inspired by the challenge, proceeded to outdo me. He drew a detailed picture of his cell, indicating precisely how his toiletries were lined up on his shelf. He outlined the routine he was forced to perform during a strip-search: "Hands through hair, flip ears, open & say ahh, finger around gums, arms up, sack up, stick up, turn around, pray for mercy, spread 'em & cough, twice." He reported on events at the jail: fights, suicide attempts, homosexual sex, and a failed escape attempt involving a pair of nail clippers and twelve bedsheets. He insisted that none of these activities involved his own participation.

He described his morning exercise routine: curling his bed mattress, then lifting a pillowcase filled with books. He provided instructions for making no-bake cookies using items available from the jail commissary—oatmeal, peanut butter, and hot-chocolate mix. He studied Spanish, he wrote, by watching the talk show *Despierta América* on television.

He learned how to fish. He carefully pulled a thread out of his mattress, tied a comb to one end, and tossed it out the gap beneath his cell's thick wooden door. He'd battle other inmates to rake in candy bars that one of them had pushed into the common area outside the cells.

Fellow inmates in the jail's maximum security wing—the "Max-men," Longo called them—became characters in his letters, including Carlos the acid-making, rap-singing Seventh-day Adventist and Dave the skinhead Wiccan vegan. At first, he said, the inmates taunted him and called him a child-killer, but soon they grew to like him. "People see who you really are after a few days," he wrote. He had heard rumors of a $500 bounty on his head at the Oregon State Penitentiary, the prison he'd likely be sent if he were found guilty. There was also a $100 prize for the first person to rape him.

Any interaction Longo had with other inmates, even through a sealed door, violated his segregation status, and he was continually reprimanded by guards for communicating. The usual way to talk was by standing on his toilet and shouting through the vents—"vental conversation," he called it, or sometimes the "vent-phone." He was able to contact people ten cells away, though of course the nine inmates between could listen in. For a more private chat, he'd lie on his stomach on the concrete floor and speak with his neighbor underneath the door.

During many of the conversations, he said, he acted as a sort of inmate therapist—aiding Dave the skinhead with his anger, and Carlos the acid maker with his negativity. The counselors available in the jail, Longo noted, were second-rate. "They might want to go back to school," he wrote. He felt he did a better job ministering to the inmates' needs.

When he was caught for these activities, he was typically punished by being kept in his cell for twenty-three hours a day, rather than the usual twenty-one—normally he spent two hours in the day room, where he could watch television or make a phone call, and one in the exercise room. Even when he was allowed out, though, he was still kept alone. Sometimes inmates teased him by singing a rendition of the song "All By Myself" through the vent-phone.

Jailhouse food was a constant topic in his letters: "chicken casserole tinted green by the peas"; "mushy spaghetti with beef pebbles"; "already-been-chewed fruit medley." He revealed the jailhouse nickname his fellow inmates bestowed upon him: Shortstop. (The opposite of Long-Go.) He once admitted that he'd illegally acquired a pen. "A fine friend," he called it, "well maybe medium, if you get my point." He responded to my weather updates with tongue-in-cheek reports of his own—"today is mild & bright w/ a temperature of about 70°, and no noticeable wind, but I'm not sure what it's like outside." He described the "Donkey Express," a method of transporting written notes from one cell to another via inmate janitors, who hid the correspondence in their broom handles.

Much of his time was spent reading. He finished seventy-three novels, he said, in his first hundred days in jail. After that he stopped counting. To combat boredom, he fiddled with the cracks in his cell window. "Some meditate in the lotus position," he wrote, "I picked at my window."

Longo even offered advice about my girlfriend. In one letter, he compared the cultivating of a new love affair with the growing of a garden. "Give it a chance," he wrote me in regard to Jill. "A real chance w/ careful planting, watering, etc."

After a while, Longo and I had become so comfortable with each other that our continued contact seemed assured. As the summer of 2002 eased into fall, with his trial still to come, I felt the time had arrived to refocus attention on my writing project. In a letter to Longo, I stated, in the most forthright terms possible, that I wanted him to continue with his life story, but that I would not be able to pay him for it.

Longo quickly responded. "I do greatly appreciate your honesty w/ regard to the potential book monies," he wrote. He said he had no plans to initiate a conversation with anyone else. He said he'd drop his demand for money and carry on with his story. "I think you understand," he wrote, "that I have committed to you."

TWENTY-ONE

A FEW MONTHS after MaryJane Baker house-sat for the Longos, a dozen members of the Golfside Congregation of Jehovah's Witnesses went on a ski trip to Mount Brighton, in eastern Michigan. Baker and Chris Longo were among them; Baker's boyfriend was not.

Longo had never been skiing before, but this didn't stop him from attempting the area's most difficult run. While trying to slow himself down, Longo smacked his face with the top of his ski pole, opening a gash above his right eye. When he reached the bottom, Baker spotted him, bleeding heavily, and brought him to the first-aid station. He was patched up and instructed to go to the hospital for stitches.

Baker offered to escort him. With her half sister, Karyn, and a friend named Deb Palmer, she took Longo to the hospital—"a glorious hour away," he wrote. Palmer drove, and Baker sat in the back, playing nurse. Longo practically forgot about the pain. "I was busy being excited at having MJ next to me," he wrote. "I was extremely drawn to her." Even so, he knew that she was "untouchable." She was nearly twenty-five; he'd just turned seventeen.

The ski trip ended without romance, but Longo and Baker soon began spending a lot more time together. Baker usually devoted Wednesdays to performing what Witnesses call "field service"—proselytizing door-to-door. Field service is typically done

in groups of four or five: a carload traveling together, two at a time
going up to a house, the rest waiting behind in the car. By this
point, Longo had completed his homeschooling and was working
at a camera shop in the Briarwood Mall. He, too, arranged to have
Wednesdays off. Most weeks, Longo and Baker and a few other
Witnesses spent all day in the same vehicle, driving the neighborhoods of Ypsilanti.

"There couldn't have been a better venue to get to know someone," wrote Longo. In the car, Wednesday after Wednesday, he
studied Baker. He noticed the way she moved her hands when she
spoke, and how easily she seemed to work with whomever she was
paired, and the deftness with which she handled strangers who
were less than pleased to find Witnesses at the door. Within her, he
felt, was an "infinite, unfluctuating kindness." He memorized
every curl and wave in her hair. He celebrated "her smile, her joy, &
her ankles." She brought alive to him, he said, the meaning of the
Bible's Song of Solomon ("You are altogether beautiful . . . you
have made my heart beat"). He wrote that he "practically wor-
shipped her."

Yet he was not blind to their differences. "She had a propensity for
seeing the worst of everything & concentrating, even worrying about
what could go wrong," he wrote. "I was the polar opposite, always
being, probably over confident, that everything would be fine." He
sensed a sadness in her. Her smile, while radiant, appeared to require a
conscious effort to maintain. She could be "strangely passive" and
"too docile" and "sometimes cool & remote." She possessed, he wrote,
"the curious amenability of a victim." And of course she was seven
years older—he was, as he expressed it, "just a kid to her womanhood."

Late in 1991 the territory lines of the Witness assemblies in
Ypsilanti were redrawn, due to an imbalance of members. Longo
learned that Baker would be attending a different Kingdom Hall.
They had known each other for a year. "My heart hurt," Longo
wrote. He quit doing field service on Wednesdays.

* * *

They didn't see each other for a while. Then, the following winter, they went skiing again on a group trip. This time, rather than trying to show off, Longo skied with Baker. For a few hours, they managed to separate from the others. Conversation flowed easily; "we seemed to bond," wrote Longo. They helped each other up when they fell. Baker mentioned, casually, that things with her boyfriend hadn't worked out. They went into the lodge and bought hot chocolates and sat by the fire. Until this day, Longo had assumed that Baker thought of him as a kind of friendly little brother. Now, he wrote, they seemed to be on "an unofficial date."

A few weeks later, in late January of 1992, Baker's friend Deb Palmer called Longo at work—he was still at the camera shop—and told him that she and Baker were coming to the mall. She said that Baker wanted to discuss something important with him.

"I became extremely flustered," wrote Longo. Perhaps this would be an official date. He retreated to the store's back room and checked himself in the mirror "umpteen times." During a break, he visited a florist and purchased a single long-stemmed red rose. Then he became paranoid that the rose was too much—maybe this wasn't a date at all; maybe they just wanted to invite him to a party. How foolish would he look then, holding a rose? He decided to hide the flower in the tiny dorm fridge in the back room, though he had to trim several inches off the stem to make it fit.

When he saw Baker approaching the store, along with Palmer, she seemed to be holding something behind her back. Maybe, thought Longo, it was something for him, so he dashed into the back room once again, took out his shortened rose, and held it behind his back. He feared that he was about to make a fool of himself, but it was too late. Baker walked in, grinning, and from behind her back she produced a gift. It was a single long-stemmed red rose.

"I was ecstatic & almost floored," wrote Longo. "But the best part was having the satisfaction of seeing her smile broaden even more after opening slightly in splendid shock at seeing what I presented from behind my back."

Palmer left them alone, and Longo spent his forty-five-minute lunch break with Baker at the mall's coffee shop. They talked, openly, for the first time. Longo admitted that he'd fantasized about her when she was house-sitting. She'd had fantasies too, she said, and these fantasies had eventually ruined the relationship she was in. They joked about Longo's frequent visits home during that week, and their obvious intent; he'd even offered to do Baker's laundry. "Like I was going to let you wash my underwear," she said.

The conversation turned serious. "I put on my spiritual hat & discussed the fact that I would only date her w/ a mind to a future marriage," wrote Longo. He asked her if their age difference bothered her.

She admitted that for a time she'd been confused, having strong feelings for someone so much younger, but as she got to know Longo better she was impressed by his maturity. Their ages didn't matter, she told him; she thought he was more mature than anyone she'd ever considered dating. She had no doubts, she said, about their compatibility.

Longo was dizzy with excitement. "I knew that this was the first forty-five minutes of the rest of my life," he wrote. "My infatuation was being given an opportunity to turn into something real & permanent." When it was time to part, they even hugged one another. Longo couldn't wait to tell everyone about his new girlfriend; his parents, he hoped, would be as happy as he was.

"My parents," wrote Longo, "blew a gasket." He was certainly not past the bloom of youth, they said, and in no way ready to begin courting. They were shocked by the age difference; they were angered that Chris had pursued such a relationship behind their backs. Joy Longo, always more temperamental than Joe, became so heated that she finally unleashed an ultimatum. She told Chris that he could either obey their rules or move out of the house.

Chris asked his mom if she was serious. She said she was. His dad nodded his assent. There was no yelling, no throwing of objects, just words at an impasse. "In those split seconds," Longo wrote, "I made up my mind & resolved that I would not let any- thing stand in the way of a future life containing MJ & I." It was January 31, 1992. Eight days earlier, Longo had turned eighteen; by his reckoning, he was an adult, he was mature, and he was making the right choice. He was going to show his parents "at any cost, & by whatever means necessary" that their opinions about him were "dead wrong."

And so, a few days later—days in which a cold silence settled over the house—Chris loaded his belongings into his Chevy and moved into the guest room at the home of his friends Peter and Debbie Estey. Joy Longo later said that it was the most upsetting day of her life. Chris wrote that it was "the only bad memory of life w/ my parents & it's the one that ended the strong relationship."

According to Chris, Baker fully supported his decision to leave home. When, she wondered, would his parents consider him past the bloom of youth? Age twenty-five, Longo said. There's no way, Baker pointed out, that they'd be able to ignore each other for so long. There was bound to be a schism sooner or later, and it was best to get it out of the way now. The worst, she said, was over.

Longo lived with the Esteys for a few weeks, then moved into an apartment with two roommates. He saw Baker every day. They ate lunch together, and went to the movies, and strolled in Gallup Park. Baker's half sister, Karyn, usually acted as chaperone—the couple, obeying Witness edicts, kept the relationship platonic. They did not even hold hands. There was, however, no shortage of what Longo described as "romantic gazes where the unmistakable thoughts jumped between us."

One evening, Baker and Longo were invited to the Esteys'

house to see a movie. They sat on the floor watching *Robin Hood,* starring Kevin Costner. They leaned against one another, shoulder to shoulder, and then, during the climactic scene in which Robin Hood and Maid Marian expressed their attraction while Bryan Adams's "(Everything I Do) I Do It for You" played on the soundtrack, the moment became "unresistably romantic." Longo whispered to Baker that this should be their song. Everything he'd do, he said, he'd do it for her. Baker turned to face him, and they continued to lean into one another, and their lips met. "I was truly, deeply in love," wrote Longo, "& I knew in that moment that we would be together forever."

In June of 1992, five months after Longo moved out of his parents' house, he and Baker and Karyn joined a large group on a bus trip to the Jehovah's Witness headquarters in Brooklyn, New York. They stayed at a Holiday Inn in Hasbrouck Heights, New Jersey. A few nights into the trip, as the group was milling around the lobby discussing where to head for dinner, Longo gave a friend twenty dollars to take Karyn along with the group and pay for her meal. He discreetly suggested to Baker that they sneak off to the hotel's restaurant. Baker agreed.

They went in and sat down. When the maître d' walked by, Longo gave him a furtive nod. A moment later, a bread basket was delivered to the table. Baker chose a roll and began buttering it. Longo looked at her, slightly dismayed. She took a bite, and then something registered in her mind. Something about the bread basket. She dropped the roll on her plate and grabbed at the basket with both hands. She brought out a small, clear box. Almost instantly, she began to cry. She opened it and took out the diamond ring, but before she could try it on, Longo grasped her hands.

"Would you dare to be my wife?" he asked.

"You know I do," she answered, and Longo slipped the ring on her finger.

TWENTY-TWO

FROM APRIL OF 2002, when Longo first called me, until the start of his trial, almost a year later, we spoke on the telephone nearly every Wednesday. Longo's calls usually arrived early in the evening, around seven my time, and we always spoke for the full hour, until we were involuntarily disconnected. It soon became habit for me to keep to myself on Wednesday nights; I'd cook an early dinner, then brew a pot of tea and retreat to my home office, upstairs.

I would ready my telephone recording device, settle into my chair, and sip my tea and wait. When Longo's call came, I'd always feel a spike of adrenaline, and as the collect-call-from-jail message played, I tried to relax my breathing before I pressed the one key on the phone to open the line.

Our conversations usually felt unforced and chatty. Longo, it seemed, was simply letting his mind wander, pleased to have a distraction from the Lincoln County Jail. We debated about our favorite American cities (New York, San Francisco, and Chicago, Longo said, were his top three); our preferred cuts of beef (prime rib for him, tenderloin for me); and the best thing about being bumped into first class on a plane flight ("The ice cream," he said). Longo revealed his feelings about the death penalty ("No one has the right to take anybody else's life"), jailhouse shaving cream ("Soap and water work a lot better"), and the particulars of his vasectomy ("It was four lady doctors. Four young, university lady doctors. I'm like, 'Wait, whoa, this is a joke, right?'").

I asked him if he kept a picture of his family hanging in his cell. "No," he said, "I'm not quite prepared for that." We discussed, at great length, the design flaws of a jailhouse commode. One time, I shook a martini as we talked, then loudly sipped on it. "I'm pretty giddy just listening to you," he said.

Longo informed me that he could do two hundred push-ups without resting, that he has always refused to eat canned fruit, and that he would order a gin gimlet or a Manhattan if there were an inmate pub. He said that one of his life's biggest regrets was not attending college. He conceded that, despite all the hours he'd put in with the Jehovah's Witnesses, he was not particularly religious. His piety, he said, "was more strong in show. It wasn't strong internally. It was a real strong devotion to putting up appearances." When I asked him what he yearned for most in jail, he said, "Besides family?" and I said, "Yeah," and he said, "I really miss being able to just go out and get a cup of coffee."

One Wednesday, early on, we worked out the official ground rules of our relationship. Mutual honesty in all matters was the chief tenet. I promised, as well, that everything we spoke or wrote about would be kept private until his trial was over. After that, we agreed, I could publish whatever I wished; nothing would ever be off the record. I made no guarantees about waiting for appeals or other legal proceedings. Longo gave me his word that he would not speak with other members of the press, and I, upon his prompting, swore that if anybody contacted me regarding his case—investigators, the media, members of MaryJane's family—I would inform Longo as swiftly as possible.

I eventually hired a typist to transcribe all our conversations. The dialogue stretches for one thousand nine hundred and forty-nine pages, across seven Kinko's-bound volumes—a mountain of prattle strewn with sporadic rivulets of unpredictable oddness. "I don't think I could talk this much to my mom," Longo told me. "I *know* I couldn't talk this much to my mom," I replied. In fact, over the period during which we regularly spoke, Longo never called anyone else, including his parents. The one time he dialed his home number, he said, he heard his father's voice, was overcome with anxiety, and hung up.

"Do you know that famous book *Tuesdays with Morrie*?" I once asked him.

"By Mitch Albom?" he said.

"Yeah."

"I had two copies of it," he told me.

"I was going to call my book *Wednesdays with Longo*," I said.

He gave me a charitable chuckle. "Actually," he said, "I've spoken with Mitch Albom." Albom is from the Detroit area, Longo added, not far from where he once lived. Longo's management jobs in the newspaper-distribution business sometimes brought him into contact with local writers.

"You should have been Mitch Albom in Mexico," I suggested.

"I guess," he said. "But I wanted somebody that nobody would— uh, I don't want to say."

"Just say it."

"Somebody that nobody would recognize. If I said I was so-and-so, I didn't want anyone to say, 'Oh, give me a break.'"

"Like Stephen King?"

"Yeah, exactly."

"It doesn't bother me," I said. "Come on, I'm full of myself but not over-full."

"Well," he said, "I'd heard of you."

"I know," I said.

"I'm not the most well-read person in the world, so that's pretty good testimony," he said. "I mean, I could have called myself Dr. Seuss."

"Theodor Geisel?" I said, attempting to one-up him. A subtext to our relationship, one we never overtly acknowledged, was this long-running intellectual skirmish. Though Longo and I sometimes downplayed it by feigning humility ("I'm not the most well-read person in the world"), each of us, I believe, felt smarter than the other, and frequently tried to prove it. "That's Dr. Seuss's real name," I explained.

"Yeah," Longo said, "I know that."

"You do?"

"I've studied a little bit about Dr. Seuss."

"Damn," I said. It was my turn to act the naïf. "I don't want to play Trivial Pursuit against you."

"He's from Wisconsin, I believe."

"I don't know," I said. "You're over my head now." (I looked it up later. Geisel was from Massachusetts.)

"There's this whole park," he continued, "that's got statues of all the Dr. Seuss characters in bronze."

"You've out-Seussed me."

"Sorry."

"I thought I'd scored a point with Theodor Geisel. Obviously not."

"Well, I had kids," he said. He fell silent for a few moments, as if his use of the past tense had stunned him. "We've been to Seussland," he added, though in a more subdued manner.

"Okay, touché," I said, towing the conversation back toward jocularity. "I give."

And on and on and on.

It all amounted to something. Gradually, over the course of weeks and months, the nature of our interactions changed. From one call to the next, it was scarcely noticeable. But if you read the entire seven-volume transcription in a single sitting (I've done it twice), you would see it's like one of those time-lapse films—a tree sprouting in the forest; a high-rise tower going up—in which a metamorphosis occurs at a pace difficult to discern from day to day.

I started to recognize what was happening on a Wednesday in early September that Longo did not call. He'd been reprimanded for communicating with other inmates, I later found out, and had lost telephone privileges for the week. That night, as I waited by the phone, a curious feeling crept over me. I'd always thought that the calls were mostly a favor to him; he had often told me that they were the highlight of his week. "I don't think I'd be as sane if it hadn't been for you," he'd said.

But when seven o'clock passed and the phone remained silent, I felt sharply disappointed. I had things I wanted to say—about my state of mind, about the status of my romance—that I wouldn't feel comfortable mentioning to anyone else. When I'd realized, for example, that I was falling in love with Jill, I had discussed this sensation with Longo first, before I'd even told Jill. "It's good to hear," Longo had said, "but scary I'm sure."

Why did I tell Longo first? For the same reasons my letters were so candid. Longo was the only person in my life I felt morally superior to, and something about this situation produced in me an unexpected openness. When it came to my *Times* debacle, I was too humiliated to talk intimately about the subject with any of my friends. Even with my parents and sister, I scarcely spoke directly of the firing; the few times it was mentioned, the conversation swiftly descended into silence until we focused on easier, ancillary subjects, like how I was going to earn a living.

Jill was a great source of solace; when I wasn't occupied with Longo, I was spending most of my time with her. But with Jill, as well, shame usually overrode any desire I had to explore the causes of my trouble at the magazine. With Longo, though, I could talk freely and candidly. Compared with the crimes he was accused of, my transgressions seemed so petty that I found myself gabbing away, poking at the roots of my behavior without hesitation or embarrassment.

Longo seemed to fully comprehend why I so badly wanted to please my *Times* editor, and why I was incapable of admitting to her that I hadn't conducted the proper interviews. He said he liked listening to my chatter. "I enjoy hearing about it because it's real-world drama," he said. "Not the worries of the criminal element that I get in here, but normal life." He was never judgmental about what I'd done, never patronizing. Frankly, he made me feel better about myself. What I'm trying to say is that when Longo didn't call, I missed him. Without our conversation, my week seemed incomplete.

I wasn't the only one captivated by Longo. As the months passed, he began including extra materials in his packages to me. Usually they were documents like FBI reports, police interviews, and legal briefs prepared by his attorneys. He also sent much of the mail he received. There were dozens of formal letters from TV stations and newspapers requesting interviews, but also a peculiar selection of correspondence from people who seemed entranced by Longo.

"Hi Christian," one of these opened. "My name is Debi. I've wanted to write you for several months now. I honestly can't tell you why I feel so compelled to do so. All I know is something is drawing me to you and I felt I should let you know. . . . I have very strong feelings about it, so I'm following up those feelings. Please don't think I'm a total freak. I'm not. I've never done ANYTHING like this before." Men, too, sent Longo tender notes. "You may be across the miles," one card read, from a gentleman in California named Joseph, "but you're close in heart."

Longo's charm even affected Jill. In a onetime exception to my Wednesday night ritual, I had her over for dinner when Longo called. Jill had never tried to dissuade me from speaking with Longo, but she wasn't fully comfortable, either, with the amount of time I was devoting to an accused murderer. "Couldn't you make friends with someone else?" she'd asked me. Soon after Longo rang, in a moment of spontaneity, I handed her the phone. They spoke for only a couple of seconds—"Nice to meet your voice," Longo told her—but a few weeks later she received a two-page letter from him.

The letter was primarily a critique of my relationship with Jill. "He's got a lot of good emotion tied up in you even if he doesn't seem to want to let it out," Longo wrote about me. "He seems to be pretty cement-stiff in some ways." He then accused me of being self-centered and a mediocre listener. He made a few blandly approving remarks—I'm apparently "an interesting guy with a lot to offer"—before concluding with this: "You two seem a very complementary match. We just need to splash him w/ a bucket of ice water—wake him up a little."

Jill was impressed by Longo's apparent forthrightness. It was fascinating, she said, to see that he wasn't trying to win her favor by saying syrupy things about me. Also, she pointed out, he'd nailed my personality precisely.

Longo's letter to Jill disturbed me. It made me fear that I'd gone too far, that I'd allowed Longo to become too involved in my life. What would happen, I wondered, if he was actually acquitted of the murders? "Come on over, when you get a chance," I'd written in a letter, playing off his assurances that he'd be found innocent. "I'll cook you up a bison roast." But did I actually want him over for dinner?

No, I realized, I did not. I didn't tell him this—my openness with him, as it turned out, had a limit. I had no idea how well, or how poorly, my relationship with Longo might end. If he were guilty, Longo clearly had hidden within him a terrifying violent streak. To have him rooting around in Jill's life was too much for me. I had a few vivid, panicked visions of Jill being stuffed into a suitcase, and I suddenly wished that I hadn't told Longo her real name or mentioned her actual job. And even if Longo were locked up for life, a clever inmate, I knew, could extend his reach well beyond a prison's walls. I asked Jill to not respond to the letter, and she agreed without protest.

But just a single personal note from Longo, a small sample of his ability to sound sincere and intelligent and kind, had altered Jill's perceptions. Before she'd read his letter, I was her only source of information about Longo—we'd spent hours discussing the nuances of his personality—and she'd believed, as I did, that he was almost certainly guilty of the murders, though we were both determined to withhold final judgment until after his trial.

Now, however, Jill said she'd begun to feel that Longo may indeed have been away from home when his family's murders occurred. She told me she wasn't able to conceive how the writer of such a considerate and funny letter could also have killed his family. Not only that, she took to quoting from his letter—"You need to be splashed with a bucket of ice water, Mr. Cement-Stiff"—in the midst of our tiffs.

Longo's most uninhibited fan club, it seemed, was the women's wing of the Lincoln County Jail. This section happened to be adjacent to the maximum-security ward, so the female inmates could sometimes see Longo through the tall, narrow window in his cell's door, and hear his voice through the vents. He received fantasizing letters, covertly delivered to his cell, on a regular basis. Longo mailed me several of them.

"Hey Chris," began one, from a woman calling herself Cotton Candy. "I lay relaxing in my hot, scented bubble bath & I smile as thoughts of you enter my mind." A page later, after a discussion of Longo's "tight, gorgeous chest," and "deep, seductive voice," the prose swiftly escalated: "I guide your hard cock deeply into my throbbing hot wet pussy." (In this same mailing, Longo mentioned an item called a "Fi-Fi"—a rubber glove wrapped in a tightly rolled towel; "the jailhouse version," he explained, "of an inflatable doll.")

Longo's response to Ms. Candy was eventually turned over to the district attorney. Never one to use blue language—"bonehead" is his most abrasive epithet—Longo wrote of "champaigne & chocolate" and "tasting each other the way it was meant to be done" and, later, "orgasmic fulfillment."

In a second letter to Cotton Candy, also acquired by prosecutors, Longo actually impersonated me again. Throughout an eleven-page note, he sprinkled sentences and paragraphs he'd lifted from the travel articles I'd mailed him. Only he pretended it was he who'd embarked on the adventures. "I am one to take risks & chances," he crowed. He added that he dreamed of continuing his explorations as soon as he was released. He'd again roam the world and express his creativity; he'd hop trains and hitch rides, unburdened of all of life's anchors.

To another female inmate, an eighteen-year-old named Brandy Fenton with whom he traded letters by stashing them in books in the law library, Longo displayed his sensitive side. "Put your sweatshirt over your pillow," he wrote, "wrap the arms over your shoulders, squeeze the pillow as tight as you can & cry as long as you want. That sweatshirt & pillow is me."

Letters also arrived from a convicted serial killer in the Oregon State Penitentiary, named Keith Jesperson. "Every letter that he sends me," said Longo, "he talks about another person that he killed." He received very little hate mail, he said, though one of MaryJane's relatives repeatedly sent him copies of the memorial program from his family's funeral.

Sometimes, Longo seemed to like jail. Being incarcerated, he wrote, "has given me [the] greatest opportunity for introspection in my life." His bail was set at $2.5 million, and he insisted that if someone were to offer him the money, he'd refuse it. "I'm grateful for the holding pattern that my life is in now," he wrote. "If I weren't in here, there'd be much more to stress about."

Other moments were not so benign. "There's tough times every day," he admitted. The words "monotony" and "loneliness" and "depression" appeared in his letters with increasing frequency. "My existence is wasted," he wrote. Nighttimes were difficult—he averaged, he wrote, no more than three hours of disjointed sleep. He had nightmares, but those he described as "easy." To end them, you just woke up. "Reality," he said, "is what's horrible." He craved human contact; he said he'd invented illnesses simply to spend time with the jailhouse nurse.

He wrote of being overwhelmed with grief after reading police reports describing his family's remains: "I sat on the thin mattress atop the concrete slab, which formed my bed, w/ my knees tucked up to my chin, hugging my legs as though I could bring some comfort to myself." Trying to work through the pain, he composed a letter to his youngest child, Madison. He longed, he wrote, "to hold you & hug you one more time, to tell you how much I love you & to show you how sorry I am that everything was cut so short." He signed it, "With Bottomless Love & Sorrow, Daddy."

Finally, in the fall of 2002, Longo prepared to issue a formal plea to the charges, which would set in motion the countdown to his trial. We'd now been in contact for six months, and Longo marked the occasion with a letter expressing his feelings about our relationship. It was a manifesto of sorts, by turns perceptive and preachy, one that I ended up reading numerous times in the following months, as the link between us grew stranger and ever more troubling.

"I feel like we've sped through the making of a good friendship," he wrote. "Firstly, enjoying learning a little about each other & our similarities, sharing some life experiences & our strange parallels. Then going through a little cooling off due to some reticense on my part, but overcoming that w/ open & honest communication, to a point where being open, I think for both of us, comes easy & without reluctance. I still feel that it's an odd sort of friendship due to its constraints, but not an impossible one. I don't think I'm so dilluted to believe that if you weren't a journalist that it would have gotten this far, or continue, but I'm not offended by that belief. We are, in a sense, using each other, but it is an amiable position, so I'm not overly concerned. . . . I want whatever comes out [in print] to be as honest as possible, whether it's good or bad, no sugar coating. I've never had a relationship built completely, 100%, on honesty. This is a good starting point for me. It's important to me. . . . I would like nothing more in my life, at this point, than to be considered as an honest & honorable person."

TWENTY-THREE

TO SEAL THEIR engagement, Longo and Baker made a pact. They vowed to be honest with each other no matter the circumstances. "We were pouring the foundation for a life together," wrote Longo, and they agreed that open, unedited communication was essential. Every evening, when they asked one another how the day went, it wasn't acceptable to simply answer "fine." They had to talk about the "woes & pros," as Longo phrased it, of their lives. "We didn't want to be an 'I'm fine' couple," he wrote.

In September of 1992, when they'd been engaged three months, Longo encountered some financial difficulties. He was earning about $9 an hour at the camera shop, and this was scarcely enough to cover the rent on his apartment and the monthly payments he owed to LeRoy's Jewelers, where he'd purchased, on credit, Baker's three-quarter-carat diamond ring. Over the course of a few days, one of his roommates said he couldn't contribute to the rent, Longo's Chevy Cavalier blew its engine, and the ring payment came due. He pawned his saxophone but was still short on funds. Baker, whose salary at the pediatric office was lower than his, had no money to spare. Longo refused, as he put it, to "crawl back to mom & dad for assistance."

While he was at work, every sale he made increased his frustration—all that cash passing through his hands on its way to the register.

Couldn't a little of it be his? Just a hundred dollars would make all the difference. One afternoon, a customer made a $108 down payment on a camcorder, but rather than putting the money in the register, Longo stashed it in his pocket. It was, he wrote, a "senseless act of spontanaety." That night, when Baker asked him how his day was, he answered, "Just fine."

In one of his letters, Longo attempted to explain his actions—why he lifted the money; why he lied to his fiancée. "I wanted to protect her from any stresses," he wrote. "I wanted her to believe that I was a stronghold, that everything w/ me was always more than okay. And I never wanted her to know to what extent I would go to make sure that everything seemed okay; or more accurately I didn't want her to think negatively of me on any level."

The day after Longo's theft, the camera shop's manager noticed an error in the records. The store was short $108. He confronted the employees, but nobody said anything. Longo was consumed by guilt, he wrote, and the next morning he confessed to the crime and returned the money. (He paid for the ring instead of rent, and was soon forced to find a cheaper apartment.) He was fired from the camera shop, and charges were pressed. Longo was later convicted of misdemeanor embezzlement and sentenced to eighty hours of community service, which he served at the Humane Society.

Worse than any punishment, Longo wrote, was telling Baker what he'd done. He sobbed his way through the story—"probably the first time I had cried in my adult life"—while Baker listened quietly. "When I was done," he wrote, "she put both arms around me & squeezed me tighter than she'd ever squeezed me before." She forgave him. He'd been stupid, she said, but she still loved him and wanted to marry him. Longo assured her that he would never do anything like that again—"anything that was not only illegal but immoral"—and reaffirmed his vow of honesty and candidness.

He also agreed to inform the elders at his Kingdom Hall of his transgressions. "I was extremely repentant," he said. The elders placed him on temporary restrictions. This seemed minor to Longo until he learned that the restrictions meant he could not marry in a Kingdom Hall. "To me," Longo wrote, "it didn't matter where or how we got married." But for Baker it was vital that her wedding ceremony be sanctioned by the Jehovah's Witnesses, with the service performed by her favorite elder, Richard Lau, who had known her since she was a child. "She could live w/ the fact that I had stolen, embezzled & lied," wrote Longo, "but the idea of not using the Kingdom Hall & Brother Lau threatened to stop us in our tracks."

Longo suggested that they delay the wedding until the restrictions were lifted. After all, they hadn't even printed invitations. Baker told Longo that she needed to be alone for a while, to think things through. Her disillusionment was palpable, and Longo was terrified. "I had blown it," he wrote. "I didn't deserve her in the first place & now she was realizing that too." He fretted all night, he says, expecting a breakup call at any moment.

The following morning, Baker drove to Longo's apartment. She knocked on the door. Through the peephole, Longo saw the "big smile that I fell in love with." Baker had apparently come to the realization that her man was more important than her church. The wedding, she declared, was still on. There would be no delay; it was hard enough as it was, she implied, to maintain their celibacy.

So they planned a wedding. They funded everything themselves; Baker's family was financially strapped, and Longo refused to ask his family for money. "I wanted to prove a point," he said. "That I was going to survive. I was going to make it big." For income, Longo took a job with Publishers Circulation Fulfillment, a company that handled national delivery of newspapers such as the *Wall Street Journal* and the *New York Times*. The money was decent, about $15 per hour, but the hours were terrible—most days, Longo worked from midnight to 7 A.M. "A professional paperboy," he called himself.

This schedule, though, allowed him to spend a good deal of time with Baker. She was still living at her mother's home, but Longo met her for breakfast and lunch every day, and saw her most evenings as well. Sometimes Baker stayed up half the night with Longo, helping deliver papers. They became, he said, each other's "best friends & sole confidants." He described himself as being in a state of bliss, and said that Baker appeared to feel similarly. "There was never a point during our engagement," he wrote, "where I felt that she wasn't 'The One.'" On her desk at work, at the pediatric office, Baker kept a dried rose— the shortened one Longo had given her on their first date, at the mall.

The only hitch was Joe and Joy Longo's continued resistance to their son's relationship. Around Christmastime of 1992, Longo's parents expressed this displeasure more forcefully. They mailed Chris a letter.

It was two pages long, signed "Mom & Dad" but written in Joy's hand. Chris called it "a rant." Joy later said it was sent out of love, and described the tone as "pleading." The letter reiterated her concerns that Chris was too selfish and immature to marry. It said, according to Chris, that if he followed through with the wedding, this "would inevitably cause hurt to others, namely MJ." Joy never questioned her son's love for Baker; she just wanted him to give the relationship more time. There's no way, she wrote, that he was ready to start a family of his own.

Chris was stung by the letter, but it only solidified his determination to prove his parents wrong. He was going to "smash their expectations to smithereens," he wrote. "I couldn't help but imagine sending my parents a letter some years down the road, consisting of one sentence—'I told you so.'"

Baker's response to the letter was even stronger. She was personally insulted. By Longo's recollection, this is how Baker expressed her feelings about Joy's letter: "What does she think, that I'm twelve years old and can't make a responsible, thought-out decision; that I would jump stupidly into something immaturely? Who does she think I am?"

Chris and MaryJane concluded that the letter wasn't worth responding to. "Neither one of us," wrote Longo, "gave any credence to the actual words or purpose." But from then on, according to Joy, the relationship between her and Baker was "always strained."

Longo, meanwhile, had few kind words to say about Baker's family. Though he hardly knew Baker's siblings, except for Karyn, this did not stop him from writing that they came "from the lower echelon of society." In regard to their marriage, he wrote, they seemed utterly indifferent.

And so, without any parental assistance, Longo and Baker worked on their wedding. For the service and reception, they rented the auditorium and cafeteria of Huron High School, Baker's alma mater. Their theme colors, they decided, would be black and white. They hired a DJ, purchased floating-candle centerpieces, borrowed table linens, and arranged for a buffet dinner. They invited two hundred and fifty people, and all but twenty said they'd come.

Longo's bachelor party, held the night before the wedding, consisted of a group of guys in his apartment eating pizza and drinking soda. There was no alcohol—only one person at the party was actually of drinking age—and no loud music. One friend did bring him a box of condoms, which was as risqué as it got. The highlight of the evening, Longo wrote, was when Baker herself came over, at 3 A.M., and they decided to exchange the gifts they'd bought each other. By a happy coincidence, they'd again made matching purchases: wristwatches. Only Baker, however, had thought to engrave her gift. "You are my everything," it said on the back.

They were married on March 13, 1993, six weeks after Longo's nineteenth birthday and six weeks before Baker's twenty-sixth. Longo's brother, Dustin, was best man; Baker's half sister, Karyn, was maid of honor. Despite Longo's restricted status within the Jehovah's Witnesses, Brother Lau agreed to marry them. Afterward,

the couple was introduced as Mr. and Mrs. Longo, and they performed a choreographed dance to "(Everything I Do) I Do It for
You."

Longo's parents even got into the spirit. Joe gave a heartfelt
congratulatory speech, and toward the end of the evening he and
Joy pulled the couple aside. They'd known that Chris and MaryJane
had spent all their money on the wedding and did not have plans
for a proper honeymoon. They handed them an envelope. Inside
were plane tickets to Jamaica, a receipt for a prepaid stay at a bed
and breakfast, and a note that said they loved them and wished
them well. The trip would mark the first time MaryJane had ever
flown on an airplane. That night, also for the first time, they slept
in the same bed, and consummated the relationship.

They were happy. Their lives progressed smoothly. They rented a
loft in the hip, regentrified Depot Town section of Ypsilanti. Longo
was fully reinstated by the church. They adopted a dog, a dalma-
tian named Pebbles. They started collecting animation celluloids
and country-style antiques. Longo moved up the corporate ladder
at Publishers Circulation Fulfillment, first to assistant manager,
then to district manager. MaryJane, to Longo's delight, required a
total of one minute and thirty-six seconds to get herself ready to go
out. ("I timed her," he wrote.)

The Jamaica honeymoon planted a travel bug in them, and
they returned to Jamaica twice more, then went to Puerto Vallarta,
Mexico, then to the Bahamas, then back to Puerto Vallarta. Church
activities occupied many of their weeknights, and they spent
almost every weekend with an older couple named Ron and Kay
Leonard, who became particularly close to MaryJane and Chris,
almost surrogate parents.

Still, there were issues. MaryJane's shyness frequently clashed
with Chris's gregariousness. "Being around others was sometimes a
struggle for MJ," Longo wrote. He found it "grinding," he said, that

his wife was seldom willing to participate in social occasions on her own. "I was the buffer, & without me," Longo wrote, "she would feel threatened somehow; highly uncomfortable." New friendships were almost impossible for her to establish. Only with the Leonards and one or two others did MaryJane appear relaxed and secure.

Her lack of self-confidence bothered him. "She would often stand back & not join in conversation," he wrote, "because she felt herself uneducated & w/o anything to offer." Worse, he said, was that she had no desire to change this—"no urges to better herself," Longo noted. "Unexplored territory made her nervous & was better left untouched."

Chris was precisely the opposite. "I wanted to learn about art & wines & languages & history," he wrote. But he felt reluctant to pursue any of these interests for fear of "leaving MJ behind" and "creating a space between us." At times, he said, he felt somewhat stifled—socially, intellectually, and culturally.

They were also in debt. Vacations sapped their income; shopping sprees sunk them. Longo bought MaryJane a dark-red Camaro and himself an SUV. He gave her Coach purses and Etienne Aigner shoes and a closetful of designer clothes. "I tried to treat her as a princess," he wrote. Soon enough, they owed money on a dozen credit cards.

Overall, though, Longo felt fortunate to have MaryJane as his wife. "There was never any doubt in my mind how much she loved me," he wrote. "Her devotion was unmatchable." The bothersome aspects of their marriage were really no more than "minor irritants"—there was nobody, he said, with whom he'd rather spend his life.

In late July of 1996, while preparing to begin their regular game of Scrabble with the Leonards, Chris and MaryJane secretly rigged the tiles. Chris went first and played the word WEE. Then MaryJane, after a calculated pause, announced that she'd be using all seven of her tiles. She placed a P and an R above the second E, then laid down G, N, A, N, and T. "MJ & I looked deviously across the counter at each other," wrote Longo. It took the Leonards a few moments, but once they saw Chris and MaryJane's faces, there were shrieks of joy. The Longos were having a baby.

TWENTY-FOUR

ON THE FIRST OF OCTOBER, 2002, just after nine o'clock in the morning, Longo made his initial appearance in the Lincoln County Courthouse. He'd been involved in previous court hearings—to petition for state-funded attorneys; to listen as prosecutors announced their death-penalty intentions—but for these events, Longo had sat in front of a video camera in a meeting room in the jail, wearing his inmate uniform. He was seen in court only as an image on a television set, and had viewed the proceedings on the meeting room's monitor.

The jail and the courthouse were next to one another; in fact, they were connected by a third-floor walkway. Longo, however, was regarded by the Lincoln County Sheriff's Office as so great a security threat that merely transferring him from one building to the other would require more than a dozen guards and strain the resources of the county's entire law-enforcement system. Longo had a legal right to appear in court every time his case was heard, but he'd waived the right for the relatively minor hearings. Now, though, it was time for him to publicly declare his guilt or innocence, and for this he needed to appear in person.

I traveled to Newport to watch. I joined a crowd of reporters and a handful of curious locals, and we were each sent through a metal detector in the courthouse basement, then directed up the

stairs, past an assembly of armed officers. By the time the courtroom doors were opened to the public, Longo was already seated at
the defense table, flanked by his lawyers.

He was wearing a nicely tailored sage-colored suit, one that
he'd saved from his Publishers Circulation Fulfillment days, along
with a beige shirt and a brown checked tie. His back was to the
spectators, and I observed his face mostly in profile, though from
what I could tell he betrayed no signs of nervousness or stress—not
a forehead wrinkle; not a tensed jaw muscle. The presence of his
attorneys seemed only to emphasize his youthful looks: Longo was
still a few months shy of his twenty-ninth birthday, while both Ken
Hadley, to his left, and Steve Krasik, on his right, were roughly double his age. Longo, busy shuffling papers and jotting notes, looked
like their eager-to-please paralegal, fresh out of college.

Judge Robert J. Huckleberry ambled into the room, and everyone jumped up. I noticed, beneath Longo's slacks, on his right calf,
the outline of an object about the size and shape of a brick. Longo
had told me about this. The device was called a Band-It. Should
Longo try and escape from the room or harm someone, the Band-
It, which was remotely controlled by a courtroom officer, would
stun him with fifty thousand volts of electricity. This was the reason he was able to appear in court without wearing manacles.

Huckleberry began the proceedings by asking Longo if he was
prepared to enter a plea to the murder charges. The question stilled
the spectator section's murmuring and note-taking, and we all
looked up to watch Longo rise from his seat again. Longo had
denied his guilt to me so often, and so emphatically, that I expected
him to announce a forceful and remonstrative "Not guilty." But he
didn't say a word. He just stood silently.

It was Hadley who ended up speaking, in his composed and
gravel-voiced manner. "Your honor," said Hadley, "at this time, the
defendant would like to stand mute to the indictment."

"Very well," said Huckleberry. He typed a few words into his desktop computer. "The court will enter a plea of not guilty to each and every charge."

And that was that. The rest of the day was devoted to various perfunctory motions filed by the defense. I was puzzled by the unusual plea, as were all the other reporters I conferred with. After court, I tracked down Hadley and introduced myself—it was the first time we'd met—and asked him what it meant to stand mute. "It's just a procedure," he said, and he didn't clarify further.

I later spoke with a few attorneys who were unaffiliated with the case. By standing mute, I learned, Longo had inserted a degree of flexibility into his defense. In essence, he had pleaded not guilty without actually saying that he wasn't guilty. This way, if he so chose, he could later change his plea—say, to not guilty by reason of insanity—without having to declare that he had misspoken the first time. It seemed a petty detail, but in a tightly contested trial, I was told, it could be the difference between whether or not he was given the death penalty.

The legal maneuver also hinted that the defense team hadn't yet decided how to proceed with the case. This was confirmed by Longo himself. I'd flown to Newport in part to witness the arraignment but chiefly to visit Longo again, which I did a couple of days after the hearing.

"It's sort of a bipolar situation," Longo said of his attorneys. "Ken is trying to save my life, and Steve is trying to get me acquitted." Hadley, he explained, was primarily concerned with avoiding the death penalty. Krasik wanted him to walk out of his trial a free man. The two strategies, Longo told me, often seemed contradictory. He knew it was ultimately up to him to decide which path to take, but the choice, he admitted, wasn't a clear one. He said he'd even considered dismissing his lawyers and representing himself.

As Longo looked at me through the glass wall and told me of his concerns, I sensed from him something I hadn't felt before: He was unsure of himself. Even his gaze seemed anxious and unsteady.

He kept curling and extending the fingers of his left hand, as if playing a private game of rock-paper-scissors. His life was literally on the line. He didn't ask me a direct question, but by the way he paused and leaned forward in his booth, as if we could put down the phones and whisper to each other through the glass, I understood that he sought my advice.

I felt as though he were coming closer to telling the truth about the murders. By not immediately rejecting Hadley's plan to mitigate his punishment rather than try for an acquittal, he hadn't confessed to anything. But he had acknowledged, however obliquely, that his innocence would not, as he'd once written me, "be obvious within a short period of time." Perhaps this new outlook had been triggered by the reality of his trial. Judge Huckleberry had announced that, barring any unexpected rulings on the motions, jury selection would commence on February 18, 2003—four and a half months away.

I wanted to help Longo, but this was an issue with potentially dire repercussions, and I knew nothing about the law. I said that he should seek an opinion from another attorney, or perhaps contact the ACLU. Longo said these were excellent ideas, though I don't think he ever followed through.

Longo seemed grateful just to have me around, even if only to listen to him vent his confusion. He said that when he'd noticed me in court—at the end of the day we'd made brief eye contact, during which he arched his eyebrows and I raised mine back—he badly wanted to come over and say hello. He realized, though, that the courtroom officers might have triggered the Band-It if he'd attempted such a move. "You know," he said, from behind the glass of the visiting-room booth, "we've never had an opportunity to shake hands."

*   *   *

Over the next few months, Longo increased the pace of his letter-writing. Envelopes from the Lincoln County Jail began arriving at my home at least once a week—a fourteen-page letter on October twentieth; twenty-five pages on the twenty-seventh; seventeen pages on Halloween.

He told me, during a Wednesday-night phone conversation, that it seemed as if he was finally growing into the role he'd assumed in Mexico. He felt like he was becoming a real writer. If he'd only made a couple of different decisions in his life, he added, and had been blessed with a little more luck, it's a career that could have been his from the start. "We were just separated at birth is all it was," he said. I laughed at this comment, we both did, but I sensed it wasn't meant entirely as a joke.

I encouraged Longo's journalistic ambitions. He was an aficionado of unusual words, so I began seeding my letters with them, along with their definitions: *xanthic; florilegia; vermiliae, shibboleth; sesquipedalian; schadenfreude*. He always made sure to slip them into his following letter. "The inmates are giving each other candy bars wrapped in xanthic paper bows," he wrote while describing the Christmas season in jail, a week after I'd told him that *xanthic* was a fancy way of referring to the color yellow.

Longo, I knew, wore his vocabulary as a form of intellectual armor, a way to show others that even as a prisoner he was still in some ways superior. Once, as punishment for communicating with inmates, he was required to compose an essay for jail officials. The topic, printed at the top of the page, was "Why It's Important to Follow the Rules." Here's the start of his second sentence: "Some, at the very mention of the word 'rules,' immediately exhibit signs of an almost intrinsic malevolence expressed with an effusive obstinacy." When I spoke with Longo about the essay, he told me, proudly, that his chief objective had been to force the jail officials "to look up some words."

I also mailed him scores of articles and short stories, anything with a prose style I thought might interest him. I included writings from Lorrie Moore, Donald Barthelme, Tim O'Brien, Raymond Carver, James Thurber, Alice Munro, George Plimpton, and (Longo's favorite) David Foster Wallace. We discussed each one—why I'd sent the piece; what was unique about it; how he could learn from it. "You're my first victim in Finkel's Correspondence Course of Eclectic Writing," I told him. I photocopied so many pages that Longo bestowed upon me the nickname Copy Boy.

Longo said that my letters stirred his "motivational juices," and that he no longer disliked writing. "I schedule time every day to write and the time flies," he told me. There were occasions, he insisted, when he skipped the opportunity to leave his cell in order to keep working. "My pencil right now is an inch and a half," he wrote in one letter. "My calluses are getting out of control." He devoted himself to his prose "whole-souled & with much pain," and added that his writing sometimes felt "more important than even my trial."

He yearned to buy a dictionary, he wrote, but said that to do so, he'd have to forgo purchasing toothpaste, deodorant, and shampoo, which would turn him into "an eccentric hippie w/ a great command of the English language." I fell for the sob story and sent him the *New Oxford American Dictionary,* hardcover edition, which cost me forty-five dollars and fifty cents.

He wrote about how hard it was to write: "It's amongst the hardest things that I've ever had to do in my life, because it's so emotional. To me it's scary to look back at my life with a detailed focus, not because it was so horrible, but because it wasn't. . . . I can't help but feel that I don't deserve to be here, to be the one to be able to reflect on these golden years. I'm writing about my life. Life was my family. My family is gone. So where does that leave me?"

He admitted that he was plagued by self-doubt. "My writing— it's horrendous," he lamented over the phone.

"Oh, stop it," I said. "You use metaphors really nicely. Some of the phrases are really writerly."

"I appreciate it," Longo said. "I'm not trying to be extremely literary." He'd studied my articles, he told me, and noted where I'd used metaphors or flowery wording. "I was a little paranoid that I wouldn't be able to do that; that my metaphors were stuck in places that you could picture anyway."

"It's always easier to take stuff out than to put something in," I said. "My advice is, if you're in doubt, throw it in."

"I don't expect for my writing to ever be novel-worthy," he said.

We had a dozen chats like this. As we discussed the construction of the sentences themselves rather than their larger meaning, the reality of the situation—that Longo was soon to go on trial for murdering his family—seemed to be transformed from almost unspeakable horror into a forum for artistic expression. This dynamic was absurd, but ideal. I wanted to know about Longo's family life in the months before the murders, and I realized that he wanted to reveal this not through some formal interview but as part of the process of exploring his abilities as a writer.

"Your writing is excellent at description and action and painting a picture, but, like a picture, it sometimes lacks that third dimension," I wrote to him. "Try and dig just a bit deeper. . . . I just want to know, Chris, what's in your heart and in your soul. If the Chris & Mike Project is to really work—to achieve what you said you wanted—a complete, and completely honest, accounting—then this is what we'll need."

And so Longo tried to tell me what was going on inside him. He wrote and he wrote. He did, indeed, dig deeper. As the date of his trial approached, he began to explore how everything in his life, slowly and inexorably, started to come apart.

TWENTY-FIVE

WHEN LONGO LEARNED that his wife was pregnant, he created a celebratory web site. It was called MrMom-to-Be.com. On the home page was a drawing by Longo of a redheaded man with a pregnant-looking belly. There were also a question-and-answer section for expectant couples, links to several baby-related sites, and a few ultrasound photos of MaryJane's womb, with labels pointing out various body parts. The site became so popular, Longo said, that *The Daily Show with Jon Stewart* contacted them for a potential on-air interview, though the show's producers eventually decided against it. ("It's impossible to determine if that really happened," said Beth Shorr, a *Daily Show* talent coordinator. "We have no record of everyone we've contacted, only those who actually appeared on the show.")

Chris and MaryJane hired a local artist to paint scenes from Peter Rabbit on the walls of their apartment's baby room, formerly Longo's office. MaryJane knitted a child-sized quilt; Longo constructed an armoire and a toy chest. They knew, well in advance, that they'd be having a boy, and Longo was so elated—he wanted "a boy to raise in the image of his father"—that he kept in his wallet an ultrasound picture of his unborn child's foot, which he showed to all his friends.

MaryJane chose to experience the birth naturally, without

painkillers. Chris stayed beside her throughout the labor process, feeding her ice chips and placing damp towels on her forehead, though he realized there wasn't much he could do. "Seeing MJ in this kind of pain," he wrote, "was like watching the replay of an airplane crash, w/ the same feeling of helplessness." Suffering aside, the delivery proceeded smoothly, and on February 28, 1997, Zachery Michael Longo was born.

Longo filled a half-dozen pages in a letter with the story of Zachery's birth. His prose was earnest and emotional. "As much love as I had for MJ," he wrote, "the view of her & our baby together showed me that I could love more, & that you really can love someone so much that it hurts. While MJ held Zack, w/ tears of her own, I trailed my finger over his little palm. His hand closed for just a second around my finger, but love swelled my heart even more."

As Longo was composing these pages, in his jail cell, he was interrupted by a pair of corrections officers. They were conducting an impromptu contraband search—a fairly common jailhouse occurrence. One of the officers was named Jacob Accurso. According to a written report on the search, Accurso observed in Longo's cell "a large quantity of hand written notes and letters on yellow paper." Accurso then "visually scanned the yellow hand written documents for evidence of jail policy violations," which include sexually explicit material, correspondence from other inmates, or anything written in code.

Longo's pages describing the birth of his son are primarily filled with joy, but if a person with a particular mindset happened to scan them, some of the phrases Longo used might line up like this: "Seeing MJ in this kind of pain . . . I now wanted it to be over with. . . . she passed the point of no return. . . . My own anguish over what had just unfolded . . . I trailed my finger over his little palm. His hand closed for just a second around my finger. . . . a weight in my chest . . . tears in my eyes . . . tightness in my throat."

This appears to be approximately how Officer Accurso read the

letter. According to the cell-search report, Accurso recalled seeing a
sentence about a baby's hand "barely being able to encircle the
author's finger" and surmised that this might be "an indication of
some impending doom." Longo's writings, Accurso deduced,
"seemed to detail the killing of Maryjane [sic] and at least one of the
Longo children."

The cell-search report was made public and picked up by the
media. Longo Notes Appear to Detail Killing was the headline of a
lengthy article in the *Oregonian*. After Longo realized what had happened—a description of birth viewed as a confession of murder—he
said he "about went through the roof." His monsterification, he
feared, was complete. No matter what he did, he'd be perceived as a
killer. Any jury, Longo knew, might feel similarly: What you look for,
you tend to see. "Every innocuous or positive aspect can & will be
twisted for the worst," he wrote. "My name is synonymous with dishonesty." A death sentence, he reasoned, was all but guaranteed.

Longo's lawyers agreed. Krasik and Hadley felt that the cell
inspection violated his constitutional protections against unreasonable searches and seizure, self-incrimination, biased juries, and
obstruction of justice. They filed a motion asking Judge Huckleberry to dismiss the case. The motion was denied. Longo remained
in jail, and continued to write.

A few months after Zachery's birth, Longo accepted another promotion from Publishers Circulation Fulfillment. His new position,
Midwest field manager, placed him on what Longo called "the
executive fast track." He traveled on business nearly every week,
overseeing the efficiency of *New York Times* delivery in a region
extending from Nebraska to upstate New York. "I was spending my
evenings in top-notch restaurants & multi-star hotels," Longo
wrote, "all of my choosing & none of which came out of pocket."

His salary was excellent. He bought himself a pool table, a car,
and a high-end suit for every day of the workweek. The family

moved out of their apartment and into a newly purchased three-bedroom home. MaryJane left her job and devoted herself to mothering Zachery. Soon she was pregnant again. "It seemed as if our prayers were answered," wrote Longo. "To me it was exactly how a proper family should be run. . . . I had followed in the footsteps of my Dad." He was twenty-three years old.

But the new job also brought unexpected changes. Over the first four years of their marriage, MaryJane and Chris had spent precisely one night away from each other, when Longo attended his great-grandmother's funeral in Iowa. Now Longo traveled three or four days a week. This lifestyle, he wrote, quickly grew monotonous. He was away when Zachery took his first steps, away when he spoke his first words. "I really started to worry about what else I had missed or was going to miss," he wrote. "The lack of physical contact had a distressing effect." He compared himself to the father in Harry Chapin's song "Cats in the Cradle," a man who was too busy working to guide his son through the difficulties of adolescence. "I kept getting on those flights at the outset of each week," he wrote, "but now I just wanted to be home."

On April 30, 1998—a day that Longo made certain he was home—MaryJane gave birth, this time sedated, to a daughter. They named her Sadie Ann. The following morning, Longo was called away on business; a newspaper-carriers strike was looming in San Francisco, and Longo's help was needed. There was no way out of it. So he left his wife, son, and newborn daughter and flew to California. He was away for two weeks. "It was at this point," wrote Longo, "that I determined that I would soon quit this job." He vowed to MaryJane that he would not miss Sadie's first steps.

He kept his word. He walked away, he said, from a promising career at Publishers Circulation Fulfillment. (The human resources department at PCF confirmed that Longo had been an executive of the company, but would not comment on the reason for his departure.) Longo took a new job selling fireplaces to the home-building

industry for a company called Fireplace & Spa. He was promised a
decent salary, though less than he was previously earning. The real
benefit of his position, he wrote, was "being able to be w/ my family
every night." Now the Longos could attend midweek services at the
Kingdom Hall as a complete family, which pleased MaryJane.
Before long, Chris was being considered for a position as a congregational elder. "Happiness in our marriage," Longo wrote, "had
reached a new peak."

In the spring of 1999, Chris and MaryJane celebrated their
sixth wedding anniversary with a vacation in northern Mexico.
They decided to drive all the way from Michigan, and they took
Zachery and Sadie with them. The long hours in the car, especially
after the kids fell asleep, gave them an opportunity to reflect on
their relationship. According to Chris, MaryJane said that she was
"exceedingly happy in life." She praised Chris for his devotion and
honesty. Except for the camera-shop incident, which occurred
when he was eighteen years old, he had fulfilled his promise of
moral integrity. Now he was twenty-five and a father of two. "I was
proud of myself," he wrote. "We had succeeded."

Motherhood, Chris noted, seemed to have imbued MaryJane
with "a positive sense of purpose." She no longer seemed as intro-
verted or fatalistic. For Chris, his career was secondary to his family. "Real life began for me when I got home from work," he wrote.
Some of the passion between MaryJane and him had faded, as happens, but they were still best friends. "She was everything to me,"
he wrote. "I needed her, & was afraid to be w/o her."

During this road trip, MaryJane discovered that she was once
again pregnant. Several months earlier Longo had scheduled a
vasectomy, but the appointment landed on the same day as Fireplace & Spa's customer-appreciation golf outing—an event at
which Longo, as a new employee, would meet many of his future
clients for the first time. It became a widespread joke among the
Longos' friends that MaryJane was pregnant because Chris chose

to play golf. Soon after the Mexico vacation, Longo rescheduled the operation, and made sure not to miss it.

Madison Jeanne Longo was born on October 19, 1999, seven weeks premature and suffering from severe respiratory problems that required a month of hospitalization. For the first time in a while, the Longos found themselves in financial straits. Health insurance covered most of Madison's care, but Longo's income from Fireplace & Spa, much of it based on commission, never came close to expectations. The family's spending habits had not changed accordingly— they hadn't changed at all—and it wasn't long before their savings were depleted and their credit cards maxed.

Just before Madison was released from the hospital, Longo was awakened one morning by loud noises in his driveway. From the window, he saw MaryJane's car, a Ford Taurus, being towed away. It had been repossessed. Until this point, MaryJane had acted as household accountant. Forced to choose which bills to cover, she had sacrificed her own car payments to keep paying Chris's car loan and the mortgage on the house.

Longo required a car for his job, so MaryJane was left at home with three young children and no transportation. There weren't any funds available to purchase even a junker car. MaryJane, wrote Longo, "was her usual emotionless self" about the situation. The two of them had their first real argument as a married couple—a "confrontation," Longo called it. Longo was upset about his wife's handling of their finances, and felt embarrassed that the neighbors may have seen their car hauled off. "I was also mad at myself," he wrote, "for not providing sufficiently to maintain our lifestyle."

The job with Fireplace & Spa, Longo realized, was not going to pan out. He was in a precarious position financially, and didn't know what to do. "With nothing higher than a high school education," he wrote, "in a town full of college grads, the likelihood of finding a job that paid the sixty-plus thousand a year that we would require was minimal."

Longo solved this problem cleverly: He started his own business. He'd learned, in the course of selling fireplaces to building contractors, that there was a huge demand for cleaning crews—people who could prepare homes after construction by shampooing the carpets, washing the windows, and generally making the places presentable for sale. One of his fireplace clients told Longo that they were desperate for such a crew. Longo phoned a friend named Joel Foster, an elder at his Kingdom Hall, and discussed the idea with him.

Foster was intrigued. The labor was nontechnical and relatively easy; the initial investment was minor; and the potential for profit was high. "Maximum yield for the minimum output" is how Longo described the idea. MaryJane, he said, fully endorsed it.

By Longo's twenty-sixth birthday, in January of 2000, Final Touch Construction Cleaning, Inc. was a full-fledged business. It was an immediate success. Within weeks, there was too much work for Longo and Foster to handle, and they hired their first employee. Soon after, they hired several more. Longo felt, he wrote, a "sense of euphoria." Everything essential to him—his wife, his children, his work, his spirituality, his morals—seemed perfectly aligned. This was the moment, Longo later admitted, that his life may have reached its high point.

The building industry is notoriously slow-paying. A month after Final Touch was launched, thousands of dollars' worth of labor had been billed, but not a single check had been collected. At the same time, Final Touch's growth, while thrilling, also required significant infusions of cash. Longo's policy was never to turn down work, but to clean homes efficiently and profitably, the business needed equipment, like forklifts and dumpsters, that neither Longo nor Foster could afford out-of-pocket. Final Touch, wrote Longo, was "rising at an alarming rate & . . . beginning to feel the effects of the lack of oxygen."

Then, in February, while Longo was driving to meet a home

builder in the Detroit suburbs, his Dodge Durango broke down. The engine had seized, and the repair estimate was more than $5,000. Longo didn't have a thousand to spare, let alone five. The family's only other car had been repossessed. He had little choice but to rent a vehicle. The Durango sat on blocks, waiting to be fixed, while Longo covered the basics: house payments, food bills, car-rental fees, and Final Touch expenses.

Longo knew that his company had astounding potential. Once the checks started rolling in and there was a steady stream of capital, he'd have the resources to clean homes for several of the major builders in Detroit. He had run the numbers over and over. He was sitting on a gold mine. Soon enough, he figured, Final Touch should be profiting $2 million a year—a million for him and a million for Foster. He'd be able to retire, set for life, by age thirty.

But Longo also realized that his company was new, its legs still wobbly, and that something as insignificant and unlucky as a blown engine could bring the whole thing down. There was no way he would allow that to happen. "I refused," he wrote, "to let a feasible business opportunity, virtually an overnight success, concede to failure just as quickly." He often worked all night, picking up nails and scrubbing floors until sunrise, then slept fewer than three hours before heading off to the next job.

The labor seemed worth it. If Final Touch folded, he might be stuck selling fireplaces for the rest of his life, stuck with his high-school education, stuck with his mid-tier salary. This was, he felt, his big chance to break out of a cycle of mediocrity. He'd already boasted to everyone at Fireplace & Spa that his business was booming. He'd bragged to his parents. He'd informed MaryJane that he was going to be named Entrepreneur of the Year.

When the Durango died, Chris told MaryJane that he'd buy her a new car. Any car she wanted. She said she dreamed of owning a minivan, maybe one of those fancy types with the television in the back. Longo guaranteed he'd fulfill that dream.

Since the day MaryJane's car had been repossessed, Chris had been in charge of the family's accounting. He'd told MaryJane about the invoices Final Touch had sent out but failed to mention that none had been paid. Their money troubles would soon be over, and Longo figured there was no reason to subject MaryJane to undue stress by admitting they were broke.

Also, he wanted her to believe that he was a brilliant businessman. In truth, he knew, the company was dangerously undercapitalized and on the brink of bankruptcy. He'd promised his wife an expensive new van when he could scarcely afford a one-week rental. He'd told his friends that the Durango breakdown was nothing more than a hiccup in his plans. He needed to think of something, quick.

On the morning of Wednesday, February 16, 2000, he came up with a plan. It was one, he thought, that might solve his problem in a couple of hours. He began by scanning his Michigan driver's license into his home computer. Using Paint Shop Pro, he erased all the data and filled in the blank spots with false information. He selected a random name out of the phone book—Jason Joseph Fortner—and a random address. The photo he left untouched.

He printed the new license and drove his rental car to an office-supply store, where he purchased a laminator. He plugged the laminator into the car's lighter-socket power converter, ran his fake license through, and cut it to the same size as his real one. The only obvious difference between the two was that the laminate on the fake license was not embossed with small holograms of the state seal of Michigan.

Longo walked back into the office-supply store and returned the laminator. He wanted to save the fifty bucks. Then he drove south, over the state line and into Ohio, to a row of car dealerships where, he hoped, the salespeople wouldn't notice the missing holograms on an out-of-state license.

He'd rented his car from Enterprise, and there was an Enterprise branch on the dealership row. It wasn't the one from which he'd hired the vehicle, but he was able to return it there anyway. This left him with no car, thereby increasing his resolve. Longo walked across the street to a Toyota dealership. This, he promised himself, in order to calm his nerves, was just a trial run. He feigned curiosity about a new car and was asked by a salesperson if he'd like to drive it. All that was necessary, he was informed, was a copy of his driver's license. No thanks, Longo replied. He said he'd return later with his wife.

He walked to an Oldsmobile dealer. There he spotted a nice minivan and mentioned to a saleswoman that he was interested in a test drive. She asked his name, and he said, "Chris." Only after he handed over his fake license for the woman to photocopy did he recall that the name on it was "Jason." He worried that he was about to be arrested. But apparently the saleswoman didn't notice, for she returned to the showroom and handed Longo his license and a car key. Longo sat in the driver's seat, and then—this was something he hadn't counted on—the saleswoman climbed into the passenger seat. They drove around for a few minutes before Longo said that he wasn't interested.

He tried again at a Pontiac dealer. In an outdoor lot, he saw a dark red Montana minivan, loaded with options, including a rear-seat video monitor. The sticker price was $34,000. A saleswoman approached, and they went through the routine of photocopying the license. This time, though, the woman came back and said, "Here you are, Mr. Fortner," and gave Longo a set of keys and a license plate. She said to put the plate in the minivan's rear window. Longo asked if it was okay to drive fifteen minutes to his wife's office, and she said, "Take all the time you need." He drove off the lot and never returned.

"When I came home w/ the van," Longo wrote, "MJ was ecstatic." It was exactly what she'd wanted, right down to the color of the leather interior. He said it was a gift to mark their seventh anniversary. She named it the Witness Wagon.

He'd stolen the vehicle, he wrote, "in order to keep up the appearance of success & to not halt the progression of our company." He convinced himself that it wasn't even a real theft—once Final Touch was flourishing, he'd be sure to send payment to the dealership and charm his way out of having any charges pressed. "No one was the wiser," he wrote. "The van issue was seamless to all but my own weighted conscience."

MaryJane, however, was somewhat skeptical. Three months previous their car had been repossessed, and now they'd bought a top-of-the-line minivan. When, she wondered, had their fortunes changed? Chris explained that Pontiac was offering a payment plan that included a ninety-day grace period, so they wouldn't owe anything until spring, by which time Final Touch would be thriving.

And what, she questioned, was the rationale for removing the license plate from the broken-down Durango and attaching it to the Montana? That was only temporary, Longo answered. He then purchased a new vanity plate for his Durango—one that read KID-VAN. When the plate was mailed to him, registered to the Durango, he simply hung it on the Montana.

Another time, MaryJane wondered where all the junk mail was. When they'd previously purchased cars, she said, there was always a flood of mail from the dealership. They'd received some, Longo told her, but he'd thrown it away. And, he added, because he'd set up an online payment plan, most of the junk mail was actually junk e-mail. Later, on his computer, he forged a document that seemed to be from Pontiac and mailed it to himself. He showed it to MaryJane. After that, he said, she had no more questions.

Final Touch continued its wild growth. Longo and Foster hired a dozen employees, then another dozen, then a third. Money finally began to trickle in, enough so that Longo could fix the Durango, and the family once again had two vehicles (one of them stolen and with a misregistered plate; the other with a canceled plate).

Best of all, Longo's dad got involved. His parents had moved back to Indiana several years earlier, and his relationship with Joe and Joy had never fully thawed. Ever since he'd left home, he'd been trying to prove that he could make it on his own. Now Joe agreed to invest several thousand dollars in Final Touch and launch the company's Indianapolis branch. This, Longo said, was one of the proudest moments of his life.

Even with Joe's help, though, the business was unraveling as fast as it was growing. Huge profits always seemed to be just around the corner. Final Touch soon had sixty employees; the payroll came to more than $15,000, due every other Friday. Joe Longo added additional money—eventually he invested a total of $60,000—but it still wasn't enough. "Elation quickly turned into frustration & stress along w/ fear," Longo wrote. "Desperation was setting in." He didn't tell his wife how he felt, and he didn't tell his father. "I would not let anyone else see the turbulence that I was going through," he wrote.

Frantic for cash, he called the builders that owed him money. They all made excuses. Foster, according to Longo, was too busy with church obligations to devote the time needed to tend to the company's troubles—sometimes he wasn't billing for a job until a month after its completion, further delaying income. "I was livid," Longo wrote. He felt as though he were doing all the work and suffering all the stress.

Many of Final Touch's employees were Longo's friends and fellow Witnesses. He couldn't ask them to delay their salaries for a few months, but he couldn't afford to pay them. He couldn't return the forklifts he'd purchased and the dumpsters he'd rented without stalling operations. And he couldn't fold the company. He couldn't even think of that. "To me it was a matter of pride & self-worth as much as anything else," he wrote. "It would be a failure, my failure."

He'd be disgraced in front of his friends, his church, his family, and his father.

With the next payday approaching and Final Touch's bank balance at zero, Longo knew that something had to give. He was prepared, he wrote, "to do just about anything to plow through the impending roadblock." First he got rid of Foster, dissolving the partnership and agreeing to pay him as a consultant. Then, the type of thinking that had come over him when he'd stolen the Montana van cropped up again. He had two days to procure $15,000, and legally or not, he was going to do it.

And finally, in the midst of all of this, his life became more complicated. Way more complicated. He fell in love with another woman.

TWENTY-SIX

IN THE MONTHS leading up to his trial, Longo met several times with a clinical psychologist named Stephen S. Scherr. Dr. Scherr, based in Portland, Oregon, was hired by Longo's attorneys in hopes that he would provide fodder for the defense. As it turned out, his evaluation was of little help. After talking with Longo at the Lincoln County Jail for fifteen hours and administering six psychological exams, including the Weschler Adult Intelligence Scale III and the Rorschach inkblot test, Scherr issued a report in which he expressed no doubts that Longo was guilty of all the murders.

Longo mailed me his copy of this report. In it, Scherr noted that Longo was clearly a smart man—"and it seemed important to him," the psychologist added, "that I know that." According to Scherr, Longo exhibited "a stronger-than-usual need for affection and attention" and "a tendency to present himself in a positive light." Longo scored extremely well on the intelligence test, above the ninety-ninth percentile in vocabulary and reading comprehension, and nearly as high on the memory sections. His vocabulary, Scherr said, was "irritatingly" good.

As for the inkblots, Longo's responses demonstrated "notable evidence of self-centeredness" but "did not show evidence of a thinking disorder or psychotic disturbance." (When I asked Longo what he'd seen in the blots, he told me, "They all looked like a factory spewing out great clouds of pollution.") "Christian's primary diagnosis," Scherr concluded, "is narcissistic personality."

According to the *Diagnostic and Statistical Manual of Mental Disorders*, fourth edition (DSM-IV), an individual with narcissistic personality disorder often has a grandiose sense of self-importance, is preoccupied with fantasies of unlimited success, and can display extreme reactivity to criticism or failure. Such a person, notes the DSM-IV, may "compare themselves favorably with famous or privileged people"—perhaps someone as privileged as, say, a reporter for the *New York Times*.

I paid to have additional psychological work performed on Longo, in an indirect way. Without informing Longo of my plans, I hired three doctors—Joe W. Dixon, a forensic psychologist and trial consultant based in North Carolina; Karen Franklin, a forensic psychologist affiliated with Alliant International University in San Francisco; and Elisabeth Young-Bruehl, a psychoanalyst on the faculty at Columbia University in New York—to study a large sample of the letters Longo had sent me. None of the three doctors was aware that I was consulting with any of the others, and none had access to Scherr's report. I told them about my book project and the crimes Longo was accused of. I sent them some background materials, in which I did not disguise Longo's biographical details but did hide his identity by blacking out his name on press clippings. I asked the psychologists for any general opinions they could glean from his letters.

Their analyses were extraordinarily similar. Dixon, Franklin, and Young-Bruehl all thought that Longo exhibited many of the behaviors associated with narcissistic personality disorder. When I asked them to speculate about why the murders took place, they all arrived at comparable scenarios. It was possible, they said, that there came a point when MaryJane—a person whose approval, Longo wrote, he "was deathly afraid of losing"—had finally had enough of Longo's behavior and threatened to leave him. An action like this may have produced what's known as a narcissistic wound, a blow to his ego so great it generated in him a murderous fury.

Scherr had addressed a related idea. "This is a man who couldn't afford to let himself get embarrassed," he observed. "He desperately needed to look good in the eyes of others." In a letter to me, Longo admitted nearly the same thing: "I feared that if MJ percieved me as a failure on any level that that would instantaneously make it a fact, & I couldn't bear that," he wrote. "I needed her to continue to look at me w/ admiration & adoration." Another time, he said that if MaryJane left him, "that would be the ultimate in embarrassment."

None of this, of course, proved that Longo committed the crimes. And even if I assumed his guilt, there was no way to determine what happened on the night the murders took place, or why, without Longo telling me himself. But this, too, was problematic. The psychologists I consulted also cautioned that I should be extremely wary when dealing with Longo, far more than I had been. People like Longo, Dixon wrote to me, are incapable of honesty. "Lying is their nature. Not just their second nature, but their nature. Beware of their snares."

Not long after I'd hired the psychologists, I received a phone call from an investigator with the Oregon Department of Justice. His name was Kerry Taylor; he was assisting the prosecution in its case against Longo. Taylor knew a little about my relationship with Longo—it seemed as if he'd been given access to the jail's visitors' log and inmate phone-call records. He had apparently researched my background as well, for he was familiar with my *Times* debacle.

"Don't get sucked in by this guy," Taylor told me. "I'm not sure he has a real grasp on the difference between truth and fiction. He just twists his stories to suit his needs."

Taylor asked for my help. He said that even if Longo had never once spoken the truth to me, my insights could be valuable. The essence of the prosecution's case, Taylor implied, was to portray Longo as a pathological liar. He wanted to know what Longo and I had talked about over the phone, and he wanted copies of the letters Longo had mailed me.

I paused for a moment. Taylor's warning, on top of the counsel I'd received from the psychologists, had rattled me. I began to fear that I'd twined myself too intricately with Longo, and had lost my sense of perspective. Still, I told Taylor I wouldn't be able to help him. I'd promised Longo that I would not share anything he told me until his trial was over. Though I had broken this pledge by showing the letters to the psychologists, I'd convinced myself that a type of doctor-client privilege was in effect.

Taylor said that my cooperation could be critical—it might make the difference between a murderer being found guilty or set free. I told Taylor that he'd put me in a bind. "I'm trying to make you feel bad enough or guilty enough to speak with me," he conceded. At that moment, I did feel guilty enough, and it took a physical effort to remain silent. "The turd," Taylor continued, "is in your pocket."

I tried to pass the turd back. I told Taylor that if he subpoenaed me—if I were legally compelled to appear at Longo's trial—then I would testify, and I'd tell the truth. Longo and I had already discussed this possibility. We'd agreed that if I absolutely had to break my silence, under threat of imprisonment, then of course I would. If I talked with prosecutors or the media under any other circumstances, then in all likelihood Longo would end our correspondence. At the mention of a subpoena, though, Taylor was quiet for a few extra beats. Then he spoke.

"Can I be blunt here?" he asked.

"You can," I said.

"You would not make a very good witness," he said.

"I wouldn't?"

"No," he said. "Because of your credibility. If I put you on the witness stand, what do you think is the first thing the defense attorney is going to bring up?"

I felt my stomach go weak. "The *New York Times* incident."

"Of course," he said. "Your lies are going to be rubbed in your face. It's a cold, hard fact. If the defense can discredit a state's witness, they will. I don't blame them. It's exactly what I'd do if I were on the other team."

I had never thought of myself as a person who'd be considered unfit to testify. To hear someone else tell me so—and for me to agree with his assessment—caught me like a sucker punch. A week after Taylor's call, I was still in a funk.

In essence, Taylor was asking me to pick a side: I could continue communicating with Longo, or I could support the prosecution. I hated being placed in such a position, but the decision really wasn't that difficult. I was, deep down, a journalist, and wasn't willing to sacrifice my story. And I had given Longo my word and did not want to betray him. So I told Taylor I'd be unable to assist him.

There was, I supposed, a third option. I could have helped the prosecution and hidden this fact from Longo—acting as a kind of double agent. Such duplicity, I'm reluctant to admit, is something I might have been good at. The West Africa article wasn't my first blatant deception. I'd lied many times: to bolster my credentials, to elicit sympathy, to make myself appear less ordinary.

I was good at lying; it was difficult to catch me. Both my parents, when I questioned them, reported that they'd never considered me a deceitful person. A little manic, yes, but not a liar. "A Ritalin child without the Ritalin," was my mom's description of my youth. My sister, too, said that she'd always thought of me as honest. Yet from a fairly young age, I'd understood that a nudge against reality—an exaggerated moment, an imaginary encounter—could make an anecdote better, smoother, and more intriguing.

I once told a touching story, in the process of flirting with a girl, about a brother of mine who had died as an infant. I had no such brother. I lied about losing my virginity—I even planted an empty condom wrapper beneath my college-dorm bed so that whoever spotted it would think I'd been having sex.

I have lied about my prowess at sports, at speaking foreign languages, at playing musical instruments. Often, I've professed to have read a book I've never opened. For a while, I told people I was Canadian. I am not. There have been occasions where I've repeated falsehoods so often—I have finished *Ulysses*, I can speak French—that I nearly hypnotized myself into believing they were true. I was also slick enough so that no one ever asked me to prove my lies.

I've lied to strangers simply because it was exciting to lie, or because I wanted to impress them. Perhaps people who've spent time on the internet pretending to be someone they're not can understand—that sense of risk, of power, of semi-illicit thrill. I *liked* lying. It could, for me, equal the escapist exhilaration of a drug.

I always thought that my journalism was immune to such impulses. I wrote creatively at times; I condensed plots and simplified complications and erased some chunks of time, but I was sure I'd always stay within the boundaries of nonfiction: Reality could be shaped and trimmed, but it could not be augmented. I had no intention of ever breaking that rule. No intention, that is, until I sat down to write my chocolate-and-slaves story.

When I returned home from West Africa to write the article, in July of 2001, I faced a tight deadline. The three previous cover-length stories I'd published in the *Times Magazine* had each taken about eight weeks to complete. For the West Africa piece, I had budgeted only about half this time. I had a rigid time limit—I was going climbing in the Himalayas with my sister. We'd been planning our expedition for two years. The trip could not be rescheduled, so I had to write quickly.

I was also shackled with some unexpected restrictions. It turned out that during the very month I was in the jungles of the Ivory Coast, so too was another *Times* reporter. His name was Norimitsu Onishi; he worked for the news department, not the magazine, but we were both chasing the same story.

The magazine and news sections are sometimes competitive with one another, and when they are, each uses its respective advantages to outdo the other—magazine articles have the luxury of length; the news section has the benefit of speed. A few days after I came home, Onishi's article ran, on the front page, accompanied by a large color photograph. To his credit, Onishi had also recognized that the slavery label had been misused. He too had walked the plantations and understood that the real issue was, as he elegantly phrased it, "the bondage of poverty." After his story was published, I thought mine would be canceled. But my editor, Ilena Silverman, felt that my piece could serve as a complement to his. "Stay away from his ideas," she told me, "and you'll be fine."

Silverman had been on maternity leave when the cocoa-plantation story was assigned, but was back in the office in time to coax me through the writing process. While she was gone, I'd teamed with another editor and had written one cover article, an investigation of the black market in human organs. I'd liked the piece, though it was, I admit, dryly written. Silverman, upon her return, informed me that she'd been somewhat disappointed: The article didn't sing the way she liked her pieces to sing; the prose wasn't rich enough.

For the West Africa article, Silverman said that it might be best to write a magazine-style feature that closely examined the journey of a single boy. Her instructions were to "go literary"—that is, as I understood it, to use a creative style to capture a reader's attention. When I realized, as I struggled through a first draft, that I wouldn't be able to provide Silverman with what she'd requested, I found myself unable to tell her. I wanted to give my editor what she wanted to read. I thought I could figure out some way to fulfill the one-boy idea and still compose a legitimate piece of journalism.

As the deadline approached and it became obvious that I'd either have to cheat or ignore my editor's instructions, I grew increasingly anxious. I wanted my story to succeed in a way that wasn't possible using the notes I'd taken. Silverman began leaving distressed messages on my voice mail—she needed to see a finished story, immediately—and I was soon in a state of panic. Eventually, as the date of my climbing trip approached, I counted the remaining hours and realized that to complete the piece, I'd have to stay awake for a long, final writing spree.

I had just the right pills. I'd gotten them for the climbing trip. Mountaineering in the Himalayas sometimes requires sleepless nights, and it is not uncommon for climbers to swallow stimulants to remain alert. I'd filled a prescription for thirty capsules of Dexedrine, ten milligrams each, a fairly potent amphetamine. The capsules were transparent, and inside them were packed hundreds of tiny, bright-orange balls. They looked like Halloween candies.

What I felt, at first, reminded me of an old playground ride—that miniature merry-go-round you'd sit on while your friends pushed you around until your vision distorted and blurred. It was like the acceleration of a centrifuge, and when I tried to concentrate on West Africa, a ream of facts and feelings and ideas were wrested from their moorings and lifted to the top of my mind. There was the color of the soil. The sounds of insects. The sting of an army ant. A sack of cocoa beans; a pair of basketball shoes. A swamp, a sunset, a skein of clouds. A cigarette, a machete, a bicycle, a bandanna.

Everything flew together and created a whole. It formed a story—what felt, as I spun, like a beautiful, flawless story; a story with passion and sadness and joy; a story of one boy that explained everything I knew to be true and yet was still a simple tale of human desire. And the boy at the center of this story, the boy composed of a hundred-and-one scattered parts, seemed to me as alive and real as anyone I'd actually met. I wrote in my journal during the experience, in black pen in a shaky hand. I described what it was like to feel dizzy and creative and manic, all at once.

"I wrote," it says in my journal, "and toiled and worked and paced and snacked and went outside to breathe and petted the cats and paced and went up and down the stairs and read my stuff and read others' stuff and wrote and wrote and wrote."

I stayed up for three days, virtually without rest, and my stomach went sour and my moods swung wildly and I puffed on marijuana when I felt out of control and popped sleeping pills to bring myself down and I never once changed my clothes and I cried without prompting and I finished the story, the entire story, and I felt it was as fine a piece of writing as I'd ever produced.

But of course the story wasn't true. Each individual piece, yes, but the whole, not at all. And I wasn't crazy. I could not plead insanity. The pressure, the time crunch, the competing story, my editor's demands, the amphetamines, the sleeping pills, and the pot are merely excuses. I knew what I was doing. I had the power to stop myself at any time, but I decided not to. It was the stupidest thing I have ever done. It's something that causes me pain every day; it's something for which I will never fully forgive myself. I wrote the story and I handed it in.

TWENTY-SEVEN

WITH FINAL TOUCH booming and, at the same time, collapsing, Longo often worked eighteen or more hours per day, attempting to supervise sixty employees across two states. Building contractors, he said, owed his company more than $100,000, but Longo was still penniless, a state of affairs he did not share with his wife, who was under the impression that much of the money had already arrived. MaryJane was likewise unaware that her new minivan had been stolen by her husband. Longo expected both issues to resolve themselves shortly—the invoices honored, the minivan paid for— but in the meantime the only thing he could count on was an ever-expanding workload.

He decided to hire an assistant. Her name was Jessica Meadows; she was the wife of another of Longo's employees, Siebert Meadows. The Meadows and the Longos lived near one another and attended the same Kingdom Hall. Jessica planned to work out of her home, and in April of 2000 Longo brought a computer over and networked it to the one in his house. Their discussion that day, however, wasn't strictly about business. According to Longo, Meadows confided in him that she was experiencing difficulties in her marriage. "I felt sorry for her," Longo wrote, "& suddenly felt an emotional attachment to her."

Over numerous lunches in the following weeks, their relationship grew. Longo became Meadows's confidant, a shoulder to cry on. "I enjoyed being in that position," he wrote, "& I began to paint a picture of myself as the knight on the white horse."

Meadows, however, remembered the conversations a little differently. She said that most of their talks were actually about how unhappy Longo was in his marriage. Longo told her that he had his own apartment where he sometimes stayed. He said that when he did come home, he always slept on the couch.

In either case, the two agreed that their connection soon deepened. "It became an infatuation of such a degree that we both began to interpret it as love," Longo wrote. One afternoon, he said, when the two of them were at lunch, he began spinning his wedding band on the table. Meadows asked him to stop; it reminded her of something her husband did. At that moment, out of the blue, Longo said that he wasn't sure if he loved his wife anymore.

"Jessica & I seemed more alike than MJ & I would ever be," he wrote. Jessica was outgoing; Jessica was carefree. He says that he thought about leaving MaryJane, and used words like "love" and "always" and "forever" with Meadows. "We both felt that it was the real thing," Longo wrote.

The relationship blew up in late May of 2000, on a day that Longo was heading to Indianapolis to organize work on a new building site. Midway through the four-hour drive from Michigan, his cell phone rang. It was Meadows. According to Longo, she was sobbing; she needed to get away for a while and she wanted to spend time with her sister in Florida, but she had no money. Longo said he'd take care of it.

He turned around. Using nearly half the remaining funds in his and MaryJane's personal account, he booked Meadows on the next available flight to Florida, early the following morning. He also stopped at an ATM to give Meadows some spending money.

They met near the airport and sat together in Longo's car, all night, waiting for her flight. "It was a clumsy passionate few hours,"

Longo wrote. They expressed some of their pent-up desires, but he insisted—as did Meadows—that their intimacy progressed no further than kissing and caressing.

At 5:30 A.M., Meadows entered the airport, and Longo headed to Indianapolis once more. He hadn't yet arrived at the building site when his cell phone rang again. This time it was MaryJane. These, according to Longo, were the first words out of her mouth: "What the hell do you have going on with Jessica?"

MaryJane, it turned out, had checked their bank balance online that morning. She noticed the recent withdrawals, some $700. Concerned, and possibly suspicious, MaryJane peeked at Chris's recent e-mails. She discovered an exchange between her husband and Meadows.

"No other woman shall ever hold the place in my affection, the way you have," said one of Longo's e-mails. "You have changed my life. It had no real meaning, no lasting laughter, and no joy. It has come back. . . . I love you and am deeply in love with you, and always will be."

In all the years Longo had known MaryJane, she'd uttered precisely two swear words in his presence. The first came after she'd been in a car accident. The second was at the beginning of this very call, when she'd used the word "hell." MaryJane now proceeded to make up for lost time in a single tirade. After she hung up on Longo, she promptly called Jessica's husband and Chris's dad and the elders at the Kingdom Hall. Then, to vent further, she drove herself and the children to Ron and Kay Leonard's house and called all her friends.

Longo left Indianapolis and rushed home—"angered," he wrote, "that so many people had been made aware of my transgression." During the trip, he made a pact with himself. He resolved to tell MaryJane the absolute truth about his affair. "She deserved honesty," he wrote. "She deserved to know where she stood & what she could expect."

He picked up MaryJane at the Leonards', where a large support group had assembled. Upon entering the house, he wrote, he felt a "hot wave of distrust that came at me from everyone in the room." He asked the Leonards to look after his kids, then drove home with MaryJane. They sat in the living room, facing one another, and talked.

Chris began, he wrote, by apologizing.

"For what?" MaryJane asked.

"Everything," he answered.

MaryJane asked if he was in love with Meadows.

"I don't know," he said.

"Are you still in love with me?" she asked.

"I don't know," he said again.

"Then what are you going to do?" MaryJane asked.

Chris said that they could try to work through the problem, if she were willing.

MaryJane said she wasn't sure. She needed to know how far he'd taken the relationship.

He recounted the complete story of the affair, and assured her that it had not been consummated. "I told her," he wrote, "that my interest in Jessica began as altruism & got carried away."

MaryJane, who had been relatively composed, now grew angry. She wanted to know what had changed in their marriage. Why, she asked, did he feel the need to look elsewhere?

There was no passion left in their relationship, Chris answered. There was no spark. They didn't have true love—they had friendship.

MaryJane began to weep. Chris tried to console her. He moved beside her and took her in his arms. He hugged her tightly, but she did not hug back.

He thought, as he grasped her, of leaving—of just escaping from his family. At that moment, he wrote, he felt as though he didn't love MaryJane, and might never love her again. He felt that maybe, for once in his life, he should do exactly what he wanted rather than what others expected. "My life," he wrote, "was always centered around pleasing and doing for someone else."

But he chose to persist. He kept holding on to MaryJane. And eventually she melted. She squeezed him back. She told him that they had to be in love, or else they'd never have made it this far—seven years of marriage, three children. She said she still loved him, even now, and couldn't bear the thought of his ever loving someone else.

Chris softened too. He told her that he could never be with another woman. He said that the affair with Meadows was over. It was, he told her, "a misperceived love." He apologized to her once more—for hurting her, for lying to her, for embarrassing her, for not living up to her expectations. He promised her complete honesty, from now on.

"She didn't forgive me in that instant," wrote Longo, "but w/in a few weeks, after seeing my attempts to correct matters, she expressed that she still believed in me & would continue to stand by me in the devotion that she had been giving me all along."

The Jehovah's Witnesses distribute a thin book, one hundred and ninety-one pages long, that many Witness couples are encouraged to read before marrying. It's called *The Secret of Family Happiness*. Longo said that he and MaryJane had both studied it, under the tutelage of a congregational elder. In chapter 3 of this book, beneath the heading "Wifely Subjugation," is an analysis of how God envisioned the proper roles of a husband and wife, as interpreted by the Witnesses.

"Marriage was not to be like a ship with two competing captains," says the book. "The husband was to exercise loving headship, and the wife was to manifest love, respect, and willing submission." A good wife, the book continues, should be "quiet and mild" in front of her husband, and should "express appreciation for his efforts in taking the lead, instead of criticizing him."

One of the people whom the prosecutorial investigator Kerry Taylor would interview in preparation for Longo's trial was a Final Touch employee named Angie McIver. Soon after Longo's relationship with Meadows ended, McIver and several other Final Touch workers ate lunch together. At this lunch, according to Taylor's report, McIver heard Longo say that he "could screw around as much as he wanted and [MaryJane] would stick around no matter what." He added that he never had to worry about her divorcing him.

Denise Thompson, the Oregon babysitter who eventually identified Zachery and Sadie's bodies, noted to another investigator that Longo never seemed to call MaryJane by her proper name. "He always referred to her as 'the wife,'" she said. Dustin Longo, Chris's brother, observed that MaryJane "would obey and follow [Chris] in whatever he wanted." Longo himself wrote that, in his family, "I was king." His wife, he added, "was loyal & tolerant to a fault."

A few days after Chris and MaryJane's living-room confrontation, Longo met with the elders at his Kingdom Hall. He was barred, provisionally, from some church activities. This, he said, actually made him feel relieved. It "left more time to be a family," he wrote. Longo also spoke with Jessica's husband, Siebert, and expressed his contrition. Siebert forgave him—the Meadows, too, had resolved to mend their relationship—and Siebert even continued working for Final Touch, though Jessica did not.

Longo drove back to Indianapolis and visited his parents. He assured them that he was correcting what he called his "temporary wayward steering." He spoke with Ron and Kay Leonard and thanked them for their support. He contacted everyone who'd come to the Leonards' house and guaranteed them all that he was now on a righteous path.

No quantity of apologies, however, could solve Final Touch's financial mess. The company had grown too big too fast. Longo had already jettisoned his partner, Joel Foster, but this did little to relieve the $15,000 payroll due every two weeks. The housing market in the Midwest had soured, and the builders were strapped, which meant subcontractors like Final Touch were paid even more sluggishly than usual. Longo wasn't able to secure a business loan, despite trying at several banks. His father had invested all he could afford. Longo's personal money had been used to fix his Dodge Durango and to purchase a twenty-five-foot boat.

Inevitably, Final Touch's paychecks began to bounce—a handful one week, more the next. The situation was dire. There appeared to be no choice except to declare bankruptcy. But Longo couldn't bring himself to do this. "At the time," he wrote, "it didn't seem like honesty was a viable option."

According to Longo, one of Final Touch's bigger builders owed him several thousand dollars. Longo had received some checks from the company, he said, but the builder was severely delinquent on many other invoices. Longo needed money immediately; there wasn't even enough time to sue the builder for the missing funds. So Longo devised another plan. He had been using the computer program QuickBooks Pro to create checks for Final Touch. He searched the program and found a template that matched the checks used by the builder. He added the company's address and account number. He scanned in their bank insignia and the signature from an original. "Other than alignment problems," Longo wrote, "it was an easy process."

On Monday, June 26, 2000, Longo walked into the National City Bank of Michigan and deposited three counterfeit checks into his account. Each check, which matched an amount on one of Final Touch's invoices, was for more than $2,500. The scam seemed to work flawlessly, so the next day Longo returned to the bank and deposited a few more. In total, he added more than $17,000 to his account. "Come Friday," Longo wrote, "I had the paychecks in hand & a smile on my face as though it were business as usual."

He wasn't worried, he wrote, that he might get caught. "All I could think of right now was that today's problem was solved." And, indeed, in the bank he'd used his real ID, made no attempt to disguise himself, and left fingerprints on the checks. "A piece of reality seemed to be missing from him," a Michigan police detective assigned to the check-fraud case later noted.

In the event that the builder discovered what he'd done, Longo had a plan. He was only taking money that was legitimately owed to him, he reasoned, and if he were confronted by the company's owners, he would simply subtract the amount of his fake checks from their tab. "I could plead w/ the builder to not press charges," he wrote, "since they were out nothing anyway."

Two weeks passed, and Longo did not hear from the company. Payroll was again due, and invoices were still not being paid, so he printed more forged checks. Then Chris and MaryJane left town for a vacation—they needed some time off, Longo said, to help recover from the Meadows affair. They took the children, camped at a lakeside park in central Michigan, and spent several days cruising on their motorboat.

Longo had purchased the boat, along with two jet skis, two forklifts, and two cargo trailers, from an acquaintance he referred to in his letters only as Travis. The equipment was so cheap—the price was a half to a tenth of retail—that Longo had suspected it might have been stolen. As it happened, everything was, but at the time Longo never questioned Travis. "I didn't want to know," he wrote. The deals were too good to pass up.

The forklifts and the trailers were important for Final Touch's operations, but the boat and the jet skis were items that Longo simply coveted. "I always had an urge to have," he admitted. He was, he conceded, both brand-aware and status-conscious in the extreme. ("I had no interest in Rolexes," he once noted. "I like TAG Heuers.") Even though both his family and his business were nearly insolvent, he couldn't resist buying luxury water toys. He told

MaryJane that he'd won the jet skis in a raffle from OfficeMax, which she seemed to believe, even though the machines were from two different manufacturers.

The weeklong camping trip, Longo said, healed his marriage. His love for MaryJane returned full force. "I became aware of just how close I came to ruining the parts of life that meant the most to me," he wrote. "That week corrected everything in my mind. . . . I would be the husband that she deserved." They were even able to joke a little about the affair. The seven-year itch, they called it.

But by the time the vacation was over, the Longos were nearly out of money. They could scarcely afford to fill the car with gas. All of their credit cards were charged to the limit. During the drive home, Longo felt desperate. So he stopped at a branch of the National City Bank. He had a few counterfeit checks with him in the minivan, ones he'd created during his original scam but had never cashed.

He left his family in the vehicle, entered the bank, and presented a check for $3,998 to a teller, along with his driver's license (his real one). The teller, Longo said, immediately appeared skeptical. She said she'd be right back, and she took his license and the check and disappeared into the rear of the bank. As Longo waited, he became increasingly paranoid. He saw employees making phone calls, people looking at him. After a few minutes, he was so nervous that he left the building, leaving behind his check and his license.

He told MaryJane that the bank was too crowded, and that he'd try again later. They played with the kids for a while, and Longo began to think that he'd overreacted, so they drove back to the bank. As soon as he walked in, though, he could sense that he was being eyed. He approached the same teller and told her that he might have accidentally left his driver's license there. He said that he was in a hurry; if she couldn't cash the check right now, he said, it was no problem. He'd do it another time. The teller said she needed to speak with a supervisor, and she disappeared again. Now

Longo was certain that something had gone wrong, and for the second time that day he fled the bank. He was panicked; his address, he realized, was printed on his driver's license.

Still, he climbed into the minivan and headed toward home. He told MaryJane nothing. As they were approaching their house, a state trooper's car that had clearly been waiting for them flipped on its lights and pulled them over. Longo was driving a stolen car and hauling a stolen boat on a stolen trailer. A police officer approached Longo's window and asked for his license, which he could not provide. The officer then asked Longo to step out of the car and place his hands behind his head.

TWENTY-EIGHT

I HAD PLEDGED to Longo that I would be completely honorable in all my dealings with him. But I failed. After the investigator Kerry Taylor phoned me, for example, I made no mention of this call to Longo, even though I'd specifically agreed to tell him of everyone who contacted me regarding his case.

I'm not exactly sure why I didn't reveal the call. After all, when Taylor had forced me to pick a side, I'd chosen Longo's. But Taylor's insistence that Longo was repeatedly lying to me had tapped a nerve, and I knew it could be risky to discuss the details of our chat. Longo might sense that I agreed with Taylor—he was expert at perceiving things like that. Longo and I had established a comfortable link, and as he neared the climax of his tale, I felt it was foolish to do anything that might jeopardize the flow of information. So I left Taylor's call unmentioned.

Not long after Taylor and I spoke, I received a phone call from a man named Carlton Smith. Smith is a writer, based in California, who specializes in true-crime books. Though Longo had refused to communicate with him, Smith was nevertheless working on a book about the Longo family murders, and also contributing articles about the case to the *Willamette Week* newspaper. (His book has since been published, under the title *Love, Daddy*.)

I'd once phoned the *Willamette Week* offices to request an issue of the paper—the printed version wasn't available in Montana, and many of the photos didn't run in the online edition. During this call I let it slip that I, too, was writing a book about Longo. Smith was immediately informed of my existence, and later contacted me.

Our conversation was fairly innocuous. We mostly chatted about the culture of Jehovah's Witnesses. I didn't share a single important tidbit of information with Smith, and he didn't impart anything to me. We did, however, briefly discuss the possibility of assisting each other sometime in the future. I'd feed Smith a bit of insider knowledge to help spice up his book, and he'd pass on some of the results of his extensive research into Longo's background. This never happened, however. Smith and I wrote our manuscripts wholly independently.

Still, I decided not to inform Longo of Smith's call. I felt guilty over the offer of information-swapping. I didn't want Longo to have the impression that I was willing to use his letters as a bargaining chip, especially after he himself, by going along with my rules for the book project, had sacrificed the opportunity to profit from his story.

When Longo phoned me the following Wednesday, he actually mentioned Smith's name early in our conversation. He said he'd received a copy of Smith's latest *Willamette Week* story. This was an ideal opening for me to reveal the chat I'd had with Smith, but I did not take it.

Fifteen minutes later, Longo again brought up Smith. He spoke about the letters Smith had written him, begging for an interview. "You should not talk to this guy," I told Longo, but I still didn't disclose Smith's call.

It was now obvious to Longo that I would never voluntarily mention it. So he raised the subject himself. "He says that you two have spoken," Longo told me. I knew, immediately, that I'd made a big mistake. His tone of voice was erased of all friendliness.

"Briefly," I responded, attempting to minimize the damage,

though already I felt the prickly sweat of shame gathering on my
forehead.

"I was waiting to see if you'd bring it up," he said. Shortly after
Smith had spoken with me, I soon learned, he'd phoned Longo's
lawyers and mentioned that he had been in contact with me. The
lawyers informed Longo. And Longo, with this information in
hand, devised a test of my trustworthiness. He wanted to see if I'd
uphold my promise and tell him of Smith's call. It was a test I
didn't come close to passing, even after Longo had given me two
deliberate nudges.

Longo asked me what Smith and I had spoken about. I gave
him a brief synopsis, but in the process I only dug myself deeper: I
omitted the part about sharing our materials.

This made Longo even more upset. "He said that there was an
offer for swapping of information," he added.

I felt like an idiot. I sensed I'd just blown every bit of trust that
Longo had invested in me. "It's possible that he got the idea that we
were going to swap something," I conceded, still trying to underplay
the problem. I insisted, though, that my offer was only in jest.
"There's no way in hell I would tell him a damn thing," I said, hoping that my pointed language would underscore some sincerity.

But I knew Longo well enough to realize that he would retaliate
in the most effective way he could, by distancing himself from me.
Possibly he'd cut off our association altogether. I wouldn't blame
him. I clutched the phone, desperate, and tried to talk my way out
of it, but I ended up tongue-tied and sputtering. Finally, I said the
only thing I could: "I'm sorry, Chris." I wasn't sure what to add, so I
apologized again.

His testiness persisted. He indicated that he might speak with
someone else. "We're still getting bugged on a biweekly basis by the
Today show," he said. "And all the national newspapers." Like that,
Longo was in control again.

I was reduced to begging him for forgiveness. I told Longo that he'd been an inspiration to me, that he'd helped me scrutinize my penchant for deception, but that I clearly hadn't yet conquered it. I asked for a second chance.

Longo said he understood what I was experiencing. "That's the same kind of loop that I went through," he said. "Both of us can say all that we want, and believability is suspect."

He wasn't ready to pardon me, though. Instead he chose this moment to tell me about his cartoon. Steve Krasik had brought an old copy of the *New Yorker* to a jailhouse meeting, and Longo had flipped through it. He'd liked a certain cartoon so much that when he returned to his cell, he drew it from memory and posted it on his wall.

The cartoon showed a cat pulling a toy car, in which was seated a mouse. Off to the side was a second mouse, clearly alarmed. He was shouting at the mouse in the car. "For God's sake, THINK! he was yelling. "Why is he being so nice to you?"

Longo said that the cartoon served as a continual reminder that he had to be wary of others' kindness. His trust in me, he said, was not merely a leap of faith but what he termed "a life leap." He said he'd taken an enormous risk by communicating with me, and there was no margin for betrayal. If I abused his trust—say, by speaking with other writers—then I could hurt his case and possibly cost him his life.

"Well," I responded, as gently as possible, "why do you think I'm being so nice to you?"

Longo wanted me to answer my own question. I had the feeling that this was another test, perhaps my final test. Until this moment, much of what I'd said to Longo had a conciliatory spin to it; a hint, perhaps, of placation. I'd told him repeatedly, for instance, that I considered him an innocent man. I'd said this, however, only because he was legally innocent, not because I thought he was actually innocent. I'd mentioned, too, that the main reason I wrote him so many letters was because he was such an appreciative and perceptive reader, as if I were generously giving of my time rather than angling for a reply. Now, though, I decided to drop all my pretenses and explain exactly why I was being so nice to him.

"It's selfishness," I said. "You're helping me work on a project." I told him that I was exploiting his tragedy for financial gain and career revitalization and personal redemption. Our friendship certainly had its genuine moments—no one had been better than him, I said, at exposing and analyzing my moral flaws—but these occasions, I admitted, were subordinate to my main goal, which was to wrest from him a story. As I spoke I felt strangely relaxed, as if I'd just given up in an arm-wrestling match.

Longo liked my confession. He liked it immensely. Right away, I could feel his anger dissipating. He said he'd been fully aware of my intentions all along, and it was good to see that I was finally speaking frankly. I was inspired by this response, so I kept talking. I told him that I'd spoken with Kerry Taylor, and I recounted to Longo the entire conversation I'd had with the investigator.

"That's good to know," Longo said, his voice warm and encouraging. "I appreciate that."

I acknowledged that I'd had three psychologists read his letters.

"I would probably do the same thing if I were in your shoes," he said, affably. "Don't be sorry. I understand."

I even told him that I didn't particularly believe in his innocence. "Who else would have done it?" I said. "All roads lead to you."

"I can't tell you how much I appreciate the honesty," Longo said. "I'd rather hear it like that than hear a buttered version."

By the time our phone call ended, Longo had forgiven me. "I'm keeping the connection with you," he declared. "It's not just a connection on the surface. I think it's deeper than that. I'm trying hard to put a lot of trust in you."

*   *   *

Up to this point in our relationship, I had never caught Longo in a lie. I had checked nearly every verifiable fact in his tale and found no solid evidence to contradict a single word he'd said to me, including his insistence of innocence. Conveniently for Longo—or, I suppose, if he were really telling the truth about the murders, extremely inconveniently—it seemed that the only people who could irrefutably confirm or deny his culpability were dead.

The start date of his trial was creeping closer. At midnight on New Year's Eve, everyone in the maximum-security wing of the Lincoln County Jail rang in 2003 by simultaneously flushing his toilet. On Super Bowl Sunday, inmates wagered push-ups and envelopes over the game's outcome. On January 23, two and a half weeks before the beginning of jury selection, Longo marked his twenty-ninth birthday.

Over this period, Longo's letters became increasingly heart-breaking and intense. He filled page after page with detailed anecdotes of his family life. He told me about going to zoos, hiking in the woods, and attending the Michigan State Fair. He recalled presents he bought his children—"a safari pop-up for Zack & an indestructable bunny book, complete w/ fur, for Sadie." He listed the names of their favorite stuffed animals: Ribbit, Bun, Raffie, Zoboomafoo. He described popping popcorn, singing lullabies, attending religious services, and jumping up and down on the green leather chair in the living room.

In one letter, he reminisced about Zachery as a newborn: "I could remember clearly how his little body felt in my arms, how he was quick to wriggle away yet loving to be held, loving me. I recalled the smells of milky breath & light soap & clean diapers & dirty ones, & I missed them all."

After four pages of this, he moved on to Sadie: "As I was gazing down she started to stretch, w/ one arm raising while the opposite leg lengthened. . . . To my joy & surprise her baby blue eyes were revealed as she looked up at me. She blinked & then she seemed to smile. It was the first time that she had communicated w/ me & I felt like bursting into song."

In another letter, he'd described taking Madison to the beach: "Maddy, every time, would bend down on a rock, smack the water w/ an open hand, be shocked by the cold splash in her face & scramble to get as far above the water as possible in mine or MJ's arms."

As I read these passages, I felt a troubling interlacing of poignancy and dread. I believed that every detail Longo mentioned was genuine. Yet at the same time I felt, as I always did with Longo, that he was giving me only part of the story, that he was carefully hiding certain pieces of the truth. The images he provided of his children were all quaint and tender and somewhat fairy-taleish—he never described a temper tantrum, never mentioned disciplining them, never struck a note of unhappiness.

Longo told me that he had composed these passages to help me with my book. It was essential, he said, to present a feel for what his family was really like. When I told him that I was attempting to double-check all the facts in his letters, Longo seemed elated—so much so that in order to assist my efforts, he mailed me a thick packet of materials that included meticulously annotated contact information for more than sixty of his relatives, friends, acquaintances, and coworkers. He also provided me with transcripts of dozens more police and FBI interviews, most of which were never made public.

All of this effort on my behalf begged a familiar question: Why was he being so nice to me? Longo's answer was that he wanted the entire story, in all its intricate detail, to be fully understood. But this response didn't feel complete.

Longo also expressed a peculiar eagerness for me to tell him of any misgivings I had about his life story, no matter how minute. "If you have any doubts," he implored, "bring them up to me." He said this repeatedly. And whenever I did raise concerns—when I asked for clarification or commented on inconsistencies or pointed out seeming lapses in logic—he was delighted. In his next letter he'd promptly patch them up. It was as though he were challenging me to find the slightest hint of dissonance in his grand saga.

And then I saw. Or at least, I thought I did. Why was he being so nice to me? For the precise reason I was being so nice to him. It was selfishness; I was helping him work on a project; he was using me.

Longo was meting out his story, a millimeter at a time, so that it could be carefully dissected. He knew I'd do a thorough job—I'd just wrecked my career and couldn't afford to make even the tiniest misstep. I'm sure Longo really did want me as a friend; I'm convinced he appreciated the lifeline. But now, I sensed, our whole relationship, almost from the very start, had been much more complex. Until this moment, I thought it had been me who'd been the cat in the cartoon. Now, I grasped, we'd both been toying with each other all along.

A disturbing feeling swept over me, an angry shock, like the moment you realize your wallet's been lifted. If the story of his family life passed my scrutiny, Longo was perhaps thinking, then surely it would pass muster with a jury. I was his dress rehearsal. I was his one-man focus group. When the time came to retell his story in court, in a matter of weeks, it would be airtight and polished, edited by his personal writing coach. What I'd unwittingly been doing, in other words, was helping Longo get away with murder.

TWENTY-NINE

AT THE POLICE STATION, under interrogation, Longo confessed to the counterfeit-check scam. The charges were serious—for each of the seven checks Longo had cashed, he could be punished with a fourteen-year prison term. Anything close to the maximum sentence was highly unlikely, though, and even the interrogating officer, Detective Fred Farkas of the Michigan State Police, attempted to allay Longo's concerns. "All things considered," said Farkas, his comments captured on the tape recorder in the interrogation room, "it's not homicide."

After Longo had been arrested, in front of their house, Mary-Jane had driven the minivan to the police station. The kids were with her. They were all waiting for him when he was released from interrogation. Longo, of course, had not told his wife about the checks—at first she'd been considered a suspect herself, but after a brief interview, it was clear to the officers that she was innocent.

Longo sat in the police station with his family and explained to MaryJane what he'd done. He insisted that he hadn't taken anything he wasn't already owed. "Although she was highly upset, she seemed to not be overly disturbed, or even surprised by the revelation of the crimes," he wrote. "After open & honest discussion w/ her, she even seemed to understand, if not agree, w/ my justifications."

While Longo was speaking with MaryJane, police officers

searched the minivan. This, said Longo, scared him deeply. Officers inspected every inch of the vehicle, confiscated his remaining counterfeit checks, and even jotted down the vehicle identification number. But they never thought to check if the van itself was stolen. And so Longo was released on his own recognizance, pending a court date.

Chris and MaryJane came to an agreement. They decided to let this incident pass as quietly as possible. They would not mention it to their friends, their families, or the elders at the Kingdom Hall. Less than a month before, Longo had been reproved by the elders for his affair with Jessica Meadows. If word of the check fraud became public, the elders could condemn him to the ultimate punishment—exclusion from the church, a process called disfellowshipping. There seemed no reason to risk such a penalty. He told MaryJane that he was already remorseful and repentant. His religious beliefs were intact. He assured his wife that he would never pursue such a course again.

By all accounts, MaryJane was an intelligent woman. One of the pediatricians she once worked for was willing to fund her medical school tuition. MaryJane didn't accept; her primary goal, it seemed, was to be a model Witness wife. Her own sister, Sally Clark, described her as "a quiet homebody" who offered "unconditional love." Another sister, Penny Dupuie, said she was "completely devoted" to Longo. Apparently, MaryJane desired more than anything to please her husband, even if a more independent woman might have cut her losses and considered a separation.

The arrest did seem to chasten Longo. He downsized Final Touch from more than sixty employees to about a dozen. He sold much of his equipment, including one of his forklifts. He gave away his jet skis to pay off his dumpster contract. MaryJane began keeping the company's books. A few invoices were paid. Longo's father, still unaware of the check frauds, remained enthusiastic about the business.

On September 21, 2000, Longo went to court. He pled guilty to four counts of fraud, but due to his fairly clean record—his only previous conviction was a misdemeanor relating to the camera-shop theft, eight years before—he received no jail time. Instead, he was required to meet with a probation officer once a month for three years, to perform eighty hours of community service, and to repay more than $30,000 to the builder in whose name he'd forged the checks. Also, Longo said, the builder never honored any of Final Touch's legitimate invoices, thereby depriving his company of additional income. Overall, though, Longo was thankful. He'd avoided going to jail and thought he would be able to keep the crimes a private matter. His plan was to forget what he called this "dark spot in our lives" and continue forward with honor, lesson learned.

The morning after Longo's conviction, MaryJane was reading the *Ann Arbor News* when she gasped and said, "Oh, no." On the second page of the local section was the headline MAN ADMITS TO COUNTER-FEITING CHECKS. Beneath that was a brief story about the crimes. They were clearly not going to be family secrets any longer. Longo had little choice but to phone the elders and arrange a meeting.

The elders disfellowshipped him. Their decision had immediate and all-encompassing consequences—most Jehovah's Witnesses cite the second book of John, verses 9 through 11 in the New World translation of the Bible, for guidance on how to treat a person who is no longer "in the teaching": "Never receive him into your homes or say a greeting to him. For he that says a greeting to him is a sharer in his wicked works."

Indeed, as soon as the announcement was made in his Kingdom Hall, no one would say a greeting to Longo. The point of such punishment, according to Longo, is for a person to see the error of his or her ways and want to return to the flock. A disfellowshipped person is still encouraged to attend services, but must sit in silence.

In a letter, Longo devoted eleven pages to describing what disfellowshipment felt like. "Where I was used to being welcomed w/ handshakes & hugs upon entering the Hall I was now invisible. Even a cold 'Hello' would have been nice, but I wasn't even acknowledged w/ eye contact, much less a word of any kind. It was strange that even the kids in the Hall avoided eye contact, all of whom I had a good rapport with, like they'd been coached by their parents. In fact they most likely were."

The punishment is not necessarily terminal, nor is it particularly rare—thousands are disfellowshipped each year, out of a worldwide population of more than six million Witnesses. With a proper display of penitence, one can be welcomed back into the congregation, sometimes in a matter of months. Often, though, the sentence is far longer. MaryJane's mom, Susan Lowery, was disfellowshipped for a decade before she was reinstated. Some never are. Longo said he'd personally known three people who'd committed suicide after being disfellowshipped.

Longo's cell phone was abruptly silenced—virtually all of his friends were Witnesses, and no one dared call him. When he was walking through town, he said, people from his congregation literally crossed the street to avoid interacting with him. His father resigned from Final Touch, demanded a return of his investment, and stopped speaking with him. His mother and brother likewise shunned him. The remaining Final Touch employees who were Witnesses also quit. "It was like being placed on another planet by myself," Longo wrote. "A complete & total obliteration of life as I had known it."

He still had MaryJane and the kids, however, whom he called the "bright & glorious light glowing at the end of a very gray tunnel." By the rules of disfellowshipment, he was no longer the head of his family; he wasn't even supposed to join them in prayer. But MaryJane, according to Longo, did not obey these edicts. "MJ was undoubtedly embarassed & still hurt, but showed me support nonetheless," he wrote. "In many ways I feel that MJ saved me from being crushed. I had everything to live for, I was still the king of my castle, I was loved & supported truly unconditionally."

MaryJane, seldom comfortable socializing on her own, essentially joined Longo in his isolation. It was too awkward for her to see their friends and pretend that everything was okay. "She seemed to shrink back into the shell of me," Longo wrote. He assured her that the situation was only temporary. He was already reformed, he pointed out, and therefore sure to be reinstated quickly. The year 2000, he agreed, had been a bad one—the affair, the check fraud, the disfellowshipment. But all they had to do was get through the winter. Come springtime, he promised, they'd once again be hosting barbecues in their backyard with all of their friends.

As it turned out, by the spring of 2001, the Longos were basically homeless. The disfellowshipment nearly killed Final Touch. The company limped on with four non-Witness employees, but Longo, desperately in need of money, also returned to delivering newspapers. He often worked all day and most of the night but still couldn't cover his bills.

In October of 2000, Chrysler had taken possession of the Durango—the only vehicle Longo legitimately owned—and was demanding $16,000 for defaulting on the lease. The family's credit-card debt was more than $30,000. Longo was required to pay $980 a month in court-assessed restitution for his forgeries. His father wanted $1,500 a month in repayment for the business loans. He owed back wages to several former Final Touch employees, a total of $12,000. He had a family of five to feed. He had a mortgage to pay. Utilities were due. He was disfellowshipped from his church, shunned by his friends, and forbidden from speaking with his parents.

Longo responded to the panic in his usual way. In January of 2001, just before his twenty-seventh birthday, he forged his father's signature on a credit-card application. The card arrived, and his family, wrote Longo, "lived w/o monetary stress for a short period." Longo decided to purchase scuba gear and diver-certification lessons for himself. He also paid for MaryJane to have laser surgery, so that her vision could be permanently corrected.

Then, one of Final Touch's remaining employees, a man named Amir Fawzy, fell off a roof while cleaning windows and broke both his ankles. Longo discovered that he'd allowed his workers' compensation insurance to lapse, so Fawzy initiated a lawsuit. That was the end of Final Touch.

When the credit line on the forged card filled up, Longo forged another. The credit-card company eventually located Longo's parents and warned them of the past-due amount. Joe and Joy agreed to help the credit-card company attempt to bring charges against their son. Between the credit cards and the unpaid loans, Longo now owed his parents more than $100,000.

In addition, he was sued by his local bank for bouncing thousands of dollars in Final Touch checks. He was sued by several former employees seeking payment. He was sued by the owners of the business that had purchased his forklift—they discovered it had been stolen and demanded their $5,000 back. He was being hounded, daily, by collection agencies. Summonses were stuck to his door. Foreclosure proceedings were initiated on his house. He was ashamed to be spotted in public—everyone in Ypsilanti, it seemed, knew he'd been disfellowshipped. His wife wouldn't even have sex with him. "Our own intimate relations," he wrote, "dwindled to nothing."

Something had to give. There was no way that Longo could repay his debts, support his family, and return to some semblance of a normal life. It was time, he wrote, to "cut our losses & try again from scratch." Though MaryJane had lived her entire life in Ypsilanti, Longo decided that they had to move.

They thought about Europe. They considered Canada. They researched Seattle and Dallas and Cincinnati and a half-dozen places in California. But then they had second thoughts about going so far away. "If we wanted to ultimately improve our lives," Longo wrote, "removal from our family & friends was not the best way to accomplish it." Also, by the terms of Longo's probation, he was not allowed to leave the state of Michigan without permission, let alone move from it. Eventually, they decided on Toledo, Ohio. It was close enough to Ypsilanti—less than an hour's drive—for Longo to attend his parole meetings, and far enough away so that the family could restart their lives in peaceful anonymity.

In order to find a place to live, Longo had to do some finagling. His credit was ruined—there was no way he could secure a mortgage or pass even a cursory background check by a potential landlord. A home or an apartment seemed an impossible goal. So instead, Longo rented a warehouse. It was a huge, hundred-and-fifty-year-old brick building, severely dilapidated, in an industrial district of downtown Toledo. The rent was $1,650 per month. It was not zoned for human occupancy.

The landlord, Pamela O'Connell, was so anxious to rent the warehouse, according to Longo, that she didn't bother with a credit check. Longo had told her that the place was going to be the headquarters for his new construction company, Urban Restoration. (The closest this idea came to fruition was the printing of a business card, on which Longo's title was "CEO.") More enticing to O'Connell, perhaps, was Longo's promise that he would provide a full year's rent as a down payment. They came to an agreement, and in May of 2001 the Longos and their dog, a husky named Kyra, left for Toledo.

Longo truly thought he'd be able to afford the warehouse. Just before they'd moved, the Longos sold their house—a step ahead of foreclosure—for their full asking price of $105,900. After paying off his mortgage loan and other fees, Longo figured he'd end up with about $16,000. He also expected a $12,000 check from the last Final Touch client.

He had, however, forgotten two things: sales tax, which reduced his house-sale profit to exactly $8,259.18, and the Amir Fawzy lawsuit, which reduced his Final Touch income to exactly zero. Longo headed to Toledo with less than a third of the nest egg he'd anticipated. He convinced the landlord to accept only one month's rent as a deposit, but to ease MaryJane's worries about the move, he told her that he'd paid rent for the whole year.

The warehouse was in such poor shape—it didn't have a kitchen or operable plumbing—that the family was forced to live in a hotel for five weeks while Longo renovated it. He purchased a water heater and a refrigerator; he rented a sand blaster and a dumpster; he paid for lumber, paint, tools, and a hotel room. "I was overextended physically & emotionally," he wrote. By the time the family moved in, on June 24, 2001, there was almost no money left.

Longo did not have a job. The family was living, illegally, in a place unfit for children—Sadie was once hit by a freefalling garage door, necessitating a trip to the emergency room. (No bones were broken.) To avoid being served with court papers, Chris and Mary-Jane told no one their street address. For mail service, they rented a box in a nearby town. They rarely turned on their cell phones; all the calls seemed to be from collection agencies. There was a Kingdom Hall not far away, but even MaryJane, whose spirituality had been a central part of her life, no longer attended services. They just needed to be by themselves for a while, Longo wrote. Funds were so tight that one afternoon he filled up the minivan with gas and drove away without paying.

MaryJane knew that they were struggling, but, Longo wrote, "she had no idea as to the magnitude of our problems." She was under the impression that there was some money left—at least enough to purchase food. There wasn't, though Longo soon changed that. He set up his computer in the warehouse and printed more counterfeit checks. He cashed five of them and netted $9,000. The checks, he wrote, were "attempts to keep us alive." This was the first time, in all the letters he'd written, that he had cast his family's misfortunes as a life-or-death battle.

*   *   *

A month and a half after they'd moved into the warehouse, the Longos had their first visitors. MaryJane's younger sister, Sally, and her husband, Anton Clark, arrived at the building. Sally was the only one of MaryJane's five siblings who was still a Jehovah's Witness. She and Anton hadn't heard from MaryJane in weeks, and had only a vague idea of where she was living. To find the place, they'd driven the streets of downtown Toledo until they spotted the Longos' dog in a fenced enclosure and their minivan in the street. Even then, they didn't see any entrance—the warehouse looked sealed off and deserted—but after they honked their horn and shouted, MaryJane eventually appeared.

The Clarks demanded to speak with Chris, so MaryJane went back into the building and brought him out. Sally and Anton told Chris that several of his former employees were still seeking unpaid wages, and were pursuing judgments against him in court. Longo assured them that things were okay. He explained that he'd already hired a lawyer to handle his affairs. "I was giving anything to pacify," Longo said, "so they wouldn't be concerned." It took him more than half an hour to persuade them to leave.

Three weeks later, on September 3, 2001, the Clarks returned. This time they came with MaryJane's mom, who was visiting from Alabama, where she lived with her husband. But when they arrived, the vehicle and the dog were gone. They managed to find the building's landlord, who told them that Longo was delinquent with the rent. He'd made several excuses, she said—that someone had broken into the warehouse and stolen his checks; that his brother had died of cancer and he had to pay for a funeral.

Alarmed, MaryJane's family tried to contact her. They dialed her cell number repeatedly, but no one ever answered. They called the police in both Michigan and Ohio. They alerted Longo's parole officer and his parents. They sent registered letters to the warehouse, but all of them came back unopened.

On September 17, they filed a missing-persons report with the Michigan State Police. The police discovered that Longo was wanted on multiple charges, including the most recent check frauds. When investigators searched the warehouse, they found photo albums, cooking supplies, children's toys, and MaryJane's wedding dress. The place was a mess. Food was still in the refrigerator. The Longos appeared to have abandoned the building in a hurry, but had left no clue where they'd gone.

What happened was this: Longo tried to sell the few valuable assets he still owned—the boat, a large-screen TV, a cargo trailer, and his one remaining forklift. He placed an advertisement in the paper and quickly sold the TV and trailer for a total of $1,000. A potential buyer contacted him about the forklift, and Longo had him come to the warehouse, where the items were stored. But when the man couldn't locate the serial numbers, he became suspicious. He told Longo he wasn't interested. Then, on his way home, he called the police.

Sergeant Paul Hickey of the Toledo Police Department arrived at the warehouse. He inspected the boat and the forklift but could not determine if either had been stolen. Longo told the officer that he didn't have the titles with him, but he'd gladly fax them to the police station the next day. Longo, Hickey later said, didn't appear nervous in the least. MaryJane was busy tending to the children. Hickey described the couple as "Ma and Pa America." There was no reason to think that their minivan was also stolen. The officer apologized for the intrusion and left the building.

MaryJane was perplexed. Longo calmed her down, he said, by insisting that the incident was some sort of misunderstanding. That afternoon, they took the children to the Toledo Zoo. On the way home, they saw two police cars and two tow trucks parked in front of the warehouse. Longo panicked. He immediately turned the car around. They went to a McDonald's and let the kids entertain themselves in the playroom for a few hours.

When they drove back to the warehouse, the police were gone—as were the boat and the forklift. Sergeant Hickey had done some research and discovered that the two items were in fact stolen. He returned to the building with the intention of arresting Longo, but no one was there. He confiscated the boat and forklift and figured he'd make the arrest later.

Hickey never got the chance. Longo admitted to MaryJane that he'd purchased the equipment under dubious circumstances and that there was a chance his ownership wasn't absolutely legal. He also said that they had to leave town right away. MaryJane didn't know about the newest round of counterfeit checks, but she was aware of his parole violations: He hadn't been paying restitution; he wasn't visiting his parole officer; he never did any community service. If the police came back, she knew he'd likely go to jail.

"MJ feared me being in jail as much as I did, for the same reasons," he wrote. "What would happen to our family?" According to Longo, MaryJane agreed with his decision to escape.

They rented a Penske moving truck. Longo promised the rental agency that he wouldn't be taking the truck out of state, so he was able to drive it away for only a $100 deposit. He packed up the truck late that night—"in a state of paranoia," he wrote, with the lights in the warehouse off, using only his flashlight to see. Mary-Jane kept the kids occupied while Longo stuffed whatever he could into the truck, leaving behind anything he deemed unimportant.

He put Sadie and Madison into the minivan, with MaryJane at the wheel, and loaded Zachery and the dog into the truck with him. They drove out of Toledo just before midnight on August 30, 2001. They'd lived there less than four months. The family, Longo calculated, had exactly $1,502 in cash. As to where they were going, Longo wrote, he knew only that he wanted to be "far enough away from everyone to just think straight."

At first, Longo was too nervous to drive on major thorough-fares—the police, he was certain, were pursuing him—so he charted a back-road course out of the state, the truck in the lead, MaryJane and the girls following behind. They made it across the state line, into Indiana, and checked into a Best Western for the night. Longo began to calm down. He convinced himself that the situation really wasn't so bad. "I didn't think of it as running away," he wrote. "It was making a change for the better."

It's hard to know if MaryJane thought the same way. She clearly understood that they were in trouble, and she knew her husband was breaking the law, but she may have believed it was her responsibility to keep her family intact. Almost certainly, she still felt protected by Chris. Three days before the Longos left the warehouse, MaryJane had met again with her sister Sally. They'd both brought their kids to the children's science center in Toledo. Sally later recalled some of what they spoke about that afternoon. She said she asked MaryJane if she ever felt unsafe around Chris, or worried that he might harm her or the children. "No, never," was MaryJane's response.

Longo himself wrote about MaryJane's reaction to the midnight escape. "I think that on the inside MJ was just as anxiety stricken as I was, but she wore a very convincing mask of indifference . . . when we finally stopped she seemed exhillerated despite the late hour & I think we both felt a sense of relief from that point on."

The next day they drove the interstates, heading west through Chicago then north into Wisconsin. While Zachery napped, Longo contemplated their situation. "I thought a lot about the stressful last year & the demise of my best laid plans," he wrote. "I kept playing the 'if only' game that never changes anything." He was immensely frightened. He could not afford to make further mistakes; he could not entangle himself with the law in any way. One slip, he understood, and he could find himself locked up for years.

He drove, and like a man in a foxhole listening for incoming bombs, he began cutting deals with himself, bargaining for his life. He swore he'd amend his ways. No more cons, no more counterfeit checks. Just honest, steady work. He'd repay his debts; he'd resolve his legal problems. There'd be no more lies. "I would get us back on track legitamately, get a job wherever we ended up, start from scratch w/ a solid foundation, both spiritually & secularly, & strive to lead a normal, modest life, w/o big dreams & aspirations."

He realized that caring for his wife and children was all that really mattered. MaryJane, he knew, was still proud of him, still supportive. She didn't care how much money he made or what kinds of status symbols they possessed. "She was saintly," he wrote. He loved her truly and deeply. The kids were spirited and healthy. Everything was going to turn around. He could sense it. From now on, he wrote, their lives would be one-hundred-percent improved.

PART FOUR

DEATH

THIRTY

SO THAT WAS HIS STORY. He never wrote a word about the day of the murders—my usefulness in helping shape his tale was evidently good only to a point. In any case, by the time Longo completed the "MJ & I Papers," as he called them, his trial was set to begin, and the crimes themselves would inevitably become the focus of attention.

It was now February of 2003. I'd been in contact with Longo for almost a year. I had written him twenty-three letters, and he had written me twenty-three letters. "I think you know more about me than my parents do," he'd told me. Yet I knew almost nothing about why his family had been killed. His letters were detailed and poignant and long, but they formed a sort of *One Thousand and One Arabian Nights,* always keeping me intrigued yet never quite reaching a conclusion. Nothing he wrote convinced me of his innocence. Still, I *wanted* to believe him. I wanted him not to be a murderer.

Maybe that was all he needed. If he could get a jury to feel the same way, perhaps he could nudge this into a form of reasonable doubt. His plan, so far as I could tell, was to use his charm as a defensive gambit. He would demonstrate that he was a bright and sensitive person; a normal, well-adjusted man. And therefore it wasn't logical that he could have committed crimes that were clearly the work of someone profoundly unhinged.

Longo did confess several misdeeds to me, but every one was weirdly altruistic, at least the way he told it. He lifted money from a camera shop to pay for MaryJane's engagement ring. He stole a minivan for his wife—rather than, say, a Ferrari for himself. He created counterfeit checks in order to pay his employees. He faked further checks to purchase building supplies that would make the warehouse safer and more comfortable for his family.

These don't seem the actions of a hard-hearted criminal. They were not vindictive crimes; they weren't especially cruel. Until the murders, Longo was never accused of a violent act. He didn't even appear to have a temper. Not once, in all our conversations, did he use a swear word. He didn't so much as raise his voice. Two of his favorite movies, he said, were *Charlotte's Web* (rated G) and *The Princess Bride* (rated PG). He claimed to have seen *Princess Bride* twenty times. He said he'd been drunk only a single time.

One woman in his Kingdom Hall in Ypsilanti described Longo as "a thoughtful, giving, helpful person." Another said he was "generous with his money and his time." A third said she "trusted him completely." A fourth said that Longo was "a model for other men" and that she'd overheard several women in the congregation say, "Why can't my husband be more like Chris Longo?"

Even after I suspected that Longo was using me to audition his testimony, I did not stop corresponding with him. I couldn't. I was immersed in my writing project; I was captivated by his tale; I was emotionally involved in his life. I did become a little more cautious about what I said to Longo, but I didn't blame him for testing his story on me. I faulted myself for not realizing it sooner.

And, I have to confess, I genuinely liked Longo. Though just about every aspect of our relationship confounded me, and though I was almost sure he had murdered his family and was lying about his innocence, I couldn't help it. As with the people in his Kingdom Hall, who still expressed their admiration for Longo even after he'd been arrested for murder, his charisma had worked on me. "I like you in a way that is beyond my control," I admitted in the last letter I wrote him before the start of his trial.

He may have killed, I thought, but there had to have been a plausible reason, some force that drove him beyond his snapping point. In the course of reading his letters, I came to believe that he really did love his family. "People can be partially bad and partially good," I wrote, "and I know for a fact that you have a good side."

I also added, in the letter I sent just prior to his trial, that the connection I shared with Longo felt "deep and profound and important." It was in no way a normal relationship, I noted, but it was real. "The sadnesses in your life, and the tragedies, and the troubles and difficulties ahead affect me too," I wrote. I told him that I wished he could erase the last few years of his life and start again. I signed the letter, "Your Friend, Mike."

"I don't know that I've ever been quite so touched—or maybe punched would be more appropo, or at least affected—by a letter," Longo responded. Our relationship, he wrote, had transformed his own life. "I can't imagine where I'd be right now, mentally & psycologically w/o the writing assignments, your help, & your person. Who would have thought that a phone call, over almost exactly a year ago now, could have transpired & progressed to this breathlessly high point?" He signed the letter, "Your Very Appreciative Friend, Chris."

Jury selection for *State of Oregon v. Christian Michael Longo,* case number 01-6441, was scheduled to begin on Tuesday, February 18, 2003. Two weeks before, Longo had told me over the phone that something surprising might happen on the very first day—something that would change the nature of the entire trial. He declined to elaborate, and though I speculated wildly, Longo simply repeated, "Just wait and see; wait and see."

The surprise actually came early. On the afternoon of Thursday, February 13, the Lincoln County district attorney's office announced that there would be a "plea hearing" in the Longo case the following day. The office released no other information.

I had known for some time that Longo's legal team, or at least Ken Hadley, was considering a plea bargain. Longo had even mailed me a copy of a letter his lawyers had sent to the district attorney's office, requesting a meeting to try and settle the case. Longo was careful to reiterate that he was absolutely innocent of the crimes—"I still have an overwhelming desire to be vindicated," he wrote me.

But, he added, because everyone assumed his guilt (the case, he conceded, was "my word against the DA's"), the trial's outcome was preordained, and spending weeks in a courtroom would only pro-long the suffering of his relatives, MaryJane's relatives, and all their past friends. Longo's parents, in their letter to the Lincoln County district attorney, wrote, "If this case goes to trial it will be like pour-ing salt in the already incredibly painful wounds."

Even if he did prove his innocence, Longo noted, "I don't know that my life would ever be 'normal' again." Therefore, he con-cluded, he'd gallantly fall on his sword and quell others' pain by going to prison for a crime he didn't commit. "I do feel an enor-mous sense of responsibility, in an indirect way, over what hap-pened," he wrote. "I don't mind living my life out incarcerated if that would be easiest for everyone else."

There were two problems with this idea. First, the district attor-ney's office had not seemed interested in bargaining. They wanted death. And second, Longo had repeatedly insisted that any agree-ment could not include his pleading guilty to murder—he wasn't that noble. Presumably, though, he'd be willing to plead to a lesser charge such as manslaughter. To me, it had seemed there was no chance the two sides would ever agree on anything.

Yet just before his trial was about to start, it appeared as if a deal had actually been made. This should not have come as a sur-prise; nationwide, most cases in which prosecutors seek the death penalty end in plea bargains. The threat of death is often used specifically to achieve such a result.

I was sure that the prosecution wasn't going to allow Longo to plead guilty to anything other than murder—the district attorney in Lincoln County is elected, and the public wouldn't stand for any softness, especially in a case involving children. What shocked me was that Longo might finally be coming clean. After all this time, he'd be confessing to the murder of his family.

Two scenarios seemed feasible. Longo would admit to all the crimes in exchange for life in prison, with or without the possibility of parole. Or he'd change his plea to not guilty by reason of insanity, and the trial would indeed go on, though in a much altered fashion. I had trouble imagining Longo, whose chief source of pride was his intelligence, agreeing to an insanity defense, but I also couldn't see anything else fulfilling his oblique hint—that the trial would continue but would be utterly changed.

As it turned out, both my ideas were wrong. Later, when I told Longo I'd thought he might plead insanity, he was insulted. "You know better than that," he said.

THIRTY-ONE

THE LINCOLN COUNTY COURTHOUSE and the Lincoln County Jail stand side by side at Newport's main intersection; two characterless structures, flanked by parking lots, that provide the town with a rather uninspiring centerpiece. From here, Newport stretches a few miles north and south, just under ten thousand people hammocked between the wide and empty Oregon beaches and the steep, green Coast Range.

Newport seems like a decent place to hide out. It's sixty miles to the nearest interstate and a three-hour drive to Portland, the closest major city. During the summer, tourism drives the local economy— the beaches, with their gothic rock formations and rolling dunes, are pristine and gorgeous—but mostly Newport is a commercial fishing community, politically conservative and not especially well-off. Homes tend to be low-slung and modest. In July, the roads are thick with RVs, but in February there are few tourists (except for the weekend of the Seafood & Wine Festival), and the town, inundated with rain, often feels somber and deserted.

The third-floor walkway that connects the courthouse and jail allowed Longo to move from his cell to his trial without having to face the public. (His reputation in Newport was encapsulated by the young man I once saw drive past the court, his car windows lowered, shouting, "Kill Longo!") For the plea hearing, Longo walked into the courtroom with his jaw set and his eyes unafraid. He wore the same sage-colored suit he'd worn at his arraignment—a finer suit than either of his lawyers wore—and his hair looked as if it had been recently barbered, with a few stray wisps spilling stylishly over his forehead. The Band-It stun device was again strapped to his right calf.

I hadn't made it to Newport for the Valentine's Day plea hearing. By the time I learned of the surprise proceeding, it was too late for me to fly or drive to Oregon. Judge Huckleberry, however, had allowed a television camera into the courtroom, and I obtained a tape of the hearing and watched it several times.

Longo himself also wrote me a detailed letter about the proceeding. He actually began the letter in the dawn hours before he was scheduled to appear in court. He described himself as "nervewrecked," and said that he'd spent much of the night pacing his cell. He wrote about his walk, escorted by a phalanx of guards, from the jail to the court. The route took him through a busy office area behind the courtrooms, and everyone turned to stare. "It was the first time I sensed that I was feared," he wrote. "I wanted to sit on the other side of their desks to ease their anxiety, to explain everything in detail."

Instead, he entered Courtroom 300. It was a compact room, cold as a cave and fluorescent-lit. The windows were covered with dark red blinds, which were never once raised during the entirety of the trial. A miniature grandfather clock hung on one wall, pendulum swinging. There were two flagpoles, one supporting the state flag of Oregon, the other a U.S. flag. The spectator section consisted of four rows of pewlike wooden benches. There were framed portraits of George Washington and Abraham Lincoln, a few shelves of thick legal books, and a calendar that read TODAY IS above a cube of white pages that were torn off daily.

On a raised platform in the far corner, occupying a tidy, three-sided work area adjacent to the witness stand, sat Judge Huckleberry, fifty-four years old, short and slightly pudgy, wearing a black robe and round, gold-framed glasses. Huckleberry was strict but not humorless. He was, I soon learned, partial to down-home platitudes— "Is that over-egging the pudding?" he'd occasionally ask if a lawyer was exaggerating a point—and while listening to testimony, he would often cradle his chin in his left hand while pinching his cheek, metronomically, with his right.

Longo sat between his attorneys. This was the second murder case for which Krasik and Hadley had partnered. The first had been a relative success—the charges were reduced to manslaughter—so they'd volunteered to collaborate again. As I watched the lawyers over the course of the trial, I realized that the dichotomy Longo had once complained about was actually an asset. In private meetings with their client, the two men tended to look at the case from different angles, which allowed for a thoughtful analysis of a range of options. In court, they presented a unified front, and their skills seemed to mesh.

It was Krasik who did most of the talking at the plea hearing. Krasik, who had been involved in many of Oregon's high-profile murder cases over the past decade, bore a passing resemblance to Albert Einstein, and possessed a wit and erudition that only enhanced the comparison. When I spoke with him over the phone a few weeks before the hearing, he'd confounded me by making references to the Dickens character Madame Defarge, a statistical concept called the five-sigma rule, and a legal stategy known as reverse Witherspooning, all within a five-minute span.

Before becoming a lawyer, Krasik had spent eleven years in the navy, where he specialized in landing jets on aircraft carriers— "Not as scary as it looks," he said, "especially if your eyes are closed." His sports jackets and slacks were often mismatched; his reading glasses were large and unhip ("They cost six dollars at Costco," he explained). Decorating his law office, I later saw, was a periodic table of chemical elements, a phrenology skull, a naval officer's sword, and a collection of Oscar Mayer wiener memorabilia. Of the fifteen death-penalty cases he'd previously defended, not one had resulted in a death sentence. "There's no such thing as a hopeless case," he once told me. "Even if the evidence is overwhelming, there's still zero and double-zero on the roulette wheel. Anything can happen."

Krasik began the plea hearing by informing Judge Huckleberry that his client no longer wanted to stand mute to the charges, as he had during his October arraignment. Longo, he added, wished to answer them himself. This is when things became very odd.

For the deaths of Zachery Longo and Sadie Longo, Krasik continued, Longo was now going to definitively plead not guilty. Judge Huckleberry nodded and confirmed that Longo had now entered his own not-guilty plea.

Then, without fanfare, Krasik said that for the deaths of Mary-Jane Longo and Madison Longo, his client would like to plead guilty. Huckleberry asked if any sort of deal had been made with the district attorney's office. No, Krasik said.

Longo was pleading guilty to a death-penalty offense without the protection of a plea agreement. This was almost unheard of; it was the legal equivalent of jumping out of a plane without a parachute. Krasik handed Longo a sheet of paper on which his guilty plea was officially spelled out. Longo signed it.

Huckleberry wanted to make sure that Longo understood what he'd just done. He addressed Longo directly. Did he realize, the judge asked, that he'd received no promises from the prosecution?

"Yes, sir," said Longo.

Did he know that pleading guilty to murder meant he was guaranteed, at the very least, a sentence of life in prison, with or without the possibility of parole?

"I do," said Longo.

Did he comprehend that, even if he was found not guilty of the

murders of Zachery and Sadie, this plea meant that he could still
be put to death?

"Yes, sir," Longo said. "I do understand that." He stayed seated
throughout the questioning. His face remained calm, but he swal-
lowed hard, and his Adam's apple bobbed.

The judge, still uncertain if Longo recognized the consequences of his actions, asked if he had consulted fully with his
lawyers about this decision.

"I have," Longo acknowledged, and the muscles in his face visibly tightened. He began to blink rapidly.

"Do you have any questions?" Huckleberry asked.

"I do not," Longo said.

Then Krasik stood up and formally read from the indictment.
He said that his client had, in December of 2001, in Lincoln
County, Oregon, caused the death of MaryJane Longo—"intention-
ally and therefore unlawfully."

Huckleberry now appeared satisfied that Longo was making an
informed, sober, voluntary decision. "Mr. Longo, I'll put it to you
directly," he said. "Is that true?"

"That's correct," Longo said. His voice sounded strained, on
the verge of breaking. His words came out almost in a gargle. His
lower lip began to tremble. "I felt like a three year old," he later
wrote me, "wilted in between the pillars of my lawyers, trying
heartily to not break down."

Krasik, still standing, continued reading. He said that his client
had, in December of 2001, caused the death of Madison Longo.
And this meant there was a further count to which Longo was
pleading guilty. In Oregon, if a murder victim is a child, a defendant can be charged with two separate counts—one for murder,
another for the murder of a child. So Longo was charged with
seven counts: four murders plus three child murders. He was pleading guilty to two of the killings and three of the counts.

Krasik now read from this last count. "The defendant," he said, "did unlawfully and intentionally cause the death of another human being, to wit: Madison Longo." He continued reading aloud. "Madison Longo," he said, "was a person under the age of fourteen years."

This was the line that finally cracked him. A single tear escaped from Longo's left eye and ran down his cheek until he swiped at it. "It was an unbearably heavy moment," he wrote. "I could only think back to that night & the fact of how guilty I was. All of the layers of shock, horror, disbelief, & shame piled on top of me all over again, and it took all I had to not scream out or collapse into a ball under the table."

The judge asked Longo to stand, and he rose wobbly to his feet, pushing aside more tears. "I'll ask the defendant this one last time," Huckleberry said. "What is your plea to the charges?"

Longo then spoke his final words of the hearing: "I'm guilty, your honor."

"Okay," said Huckleberry. "I'll find you guilty."

THIRTY-TWO

SOON AFTER RETURNING TO JAIL, through the walkway, Longo was allowed to spend a little time in the day room, where he watched the television news reports of his guilty pleas. Then, back in his cell, he continued the letter he'd begun early that morning. "No one got it. No one understood what I was doing," Longo wrote. "No one could conceive that I was simply taking responsibility & admitting my guilt."

He was correct. Here is the first line of the next morning's Portland *Oregonian* story on the hearing: "In a legal maneuver that baffled observers across the region . . ." The Newport *News-Times*: "In a move that has confounded many . . ." The Eugene *Register-Guard*: "In a maneuver with no apparent legal rhyme or reason . . ."

And here is what Longo told me: "I'm extremely sorry for the surprise that was thrust upon you. . . . But I'm ever sorrier for misleading you into believing that I felt that I was completely innocent. For lying to you. (It's hard to write the word 'lying.') I have tried to be as honest as possible, in fact setting records in my level of honesty. I know that I've been excrutiatingly open & have not buttered anything up to make it sound better or to make it more palatable."

Longo had finally admitted that he was a murderer. I'd anticipated this moment for months, envisioning what it would be like for him, and for our relationship, when he was able to drop all his layers of deception and come to terms with who he really was. But now I couldn't shake the feeling that his guilty pleas were just the beginning of another complicated game. He hadn't come clean in the least. There'd been no breakthrough. How could Longo commit two of the murders, I wanted to know, but not the other two? If he didn't kill Zachery and Sadie, then who did?

When I pressed him to answer these questions, Longo was evasive. The pleas, he stated, were a "weight off my shoulders" and "a huge release." I requested more specifics, but he said he couldn't help me right now. He asked for my patience and promised that everything would soon be clear. "My conscience will be free," he insisted, "even if my body is not."

For days, I obsessed about the meaning of his guilty pleas. I consulted lawyers (not Longo's—they were keeping silent); I researched past death penalty cases; I spoke with journalists who specialized in legal affairs. Nothing shed much light. It was baffling. One of the more popular hypotheses, in fact, was that the whole point of the plea hearing was not to have a point. It was staged precisely to confuse people, to increase the likelihood that an error would be made during the trial, to kink the roulette wheel in a way that would produce an unnatural harvest of zeros and double-zeros.

But this idea, even in a case as discouraging as Longo's, seemed absurdly risky. Its chances of hurting him far outweighed the potential to help. Longo's confession had eliminated what lawyers sometimes call "residual doubt," which is the thin but distinct gap between a jury's finding someone guilty "beyond a reasonable doubt"—the necessary standard for conviction in a criminal case—and finding someone guilty with absolute certainty. Residual doubt is often cited as the reason juries or judges find a defendant guilty but don't impose the maximum sentence.

By pleading guilty to two murders, Longo had removed this

buffer. He'd all but invited a death penalty. Now no one would worry that an innocent person had been condemned. Longo was, without question, a murderer. It seemed certain that he would never again be a free man. If Longo had admitted to all the murders, rather than half of them, he would have exposed himself to no further punishment, beyond the purely theoretical. The state can't put someone to death more than once; an inmate can't serve more than one life term.

So why didn't Longo admit to all four? The only explanation that made sense was also, confoundingly, the least plausible: Longo had pleaded guilty to only two murders because those were the only two murders he'd committed.

It took a couple of weeks to seat a jury. The selection process in a death-penalty case is different from that of any other criminal trial. In Oregon, as in all states, in a death-penalty case it is the jury, not the judge, that determines the sentence. Longo's jury needed to be "death qualified"; that is, every member had to be willing, at least in principle, to administer a death sentence. Anyone who was morally opposed to capital punishment, or could not envision authorizing another's death, was disqualified from consideration.

A capital case is divided into two parts. In the first, known as the guilt phase, the jury determines the defendant's culpability. Because Longo had already confessed to the murders of MaryJane and Madison, the guilt phase of his trial would be concerned only with Longo's role in Zachery and Sadie's deaths. Whenever I used the term "guilt phase" on the telephone—through most of his trial, we continued to regularly speak on phone—Longo expressed his displeasure. "Why isn't it called the innocence phase?" he wondered.

In the second part of the trial, the penalty phase, the jury listens to what is known as mitigating evidence—testimony, presented by the defense, about the defendant's character that might justify a less severe sentence. The prosecution, in turn, can offer aggravating evidence that argues for a stiffer penalty. Normally, if a jury finds the defendant not guilty during the first part of a trial, there is no penalty phase. But Longo's case was certain to have both parts; even if he was acquitted in the guilt phase, he would still face punishment for his role in MaryJane and Madison's murders.

For each count on which Longo was found guilty, his jury would have three choices: a life sentence with the possibility of parole after thirty years; life without the possibility of parole; or death by lethal injection. To impose death, all twelve jurors had to vote in favor. For the chance of parole, ten or more jurors had to agree. Any other configuration of votes would result in the default sentence—life in prison without any prospect of release.

All that was needed to spare Longo's life was one sympathetic juror. Both the defense and the prosecution teams were well aware of this, and during the selection proceedings, potential jurors were asked to precisely define their stance on capital punishment, and to complete a sixty-six-question survey that requested information as far-ranging as the types of firearms they owned, how they felt about the psychiatric profession, and the wording of any bumper stickers on their vehicles.

Ken Hadley took the reins for the defense's part in the jury selection. Hadley radiated calmness and civility; with him there were no courtroom theatrics, no angry outbursts, not so much as a quickly uttered sentence. During the trial, he sometimes addressed the jury while standing behind a lectern. He was silver-haired, barrel-chested, and a firm supporter, he told me, of the Republican Party, except when it came to death-penalty issues. "An ole country lawyer" is how he described himself.

Hadley was a well-known figure in Newport; his office, in a low-ceilinged annex behind an insurance agency, was an easy walk from the Lincoln County Courthouse. Before accepting Longo's case, he'd defended twenty other people facing the death penalty. Six had gone to trial, and only one was actually sentenced to death, and that was after the defendant had ignored Hadley's advice and turned down a plea bargain. "I think of him as a grandfather figure," Longo said of Hadley. "But I've told him he's a father figure, to not hurt his feelings."

Longo was present in the courtroom for jury selection, as was I. Newport had become my temporary home; I'd taken a short-term lease on a rental cottage near the ocean, and I had driven my pickup to Oregon. I spent almost every weekday in the same room with Longo and was forced to continue my romance with Jill via the telephone. "What could be healthier for our relationship," she teased, "than for you to spend a couple of months watching a murder trial?"

On the second day of the selection proceedings, as I sat in court, I was approached by an officer and served with two subpoenas. Kerry Taylor, it turned out, may have been bluffing when he'd attempted to convince me to speak with him before the trial began. He had said that I would not make a good witness, but now the prosecution had subpoenaed me once to testify in court, and once to hand over all of Longo's letters and my tape-recordings of our phone calls. Oregon, however, has a strong journalists' shield law—this may be why Taylor tried at first to cajole me out of the information—and after I hired an attorney, the two subpoenas were swiftly dropped.

As the jury was picked, Longo wrote me a letter, commenting in his meticulous way on the people who were auditioning for the job of deciding whether he should live or die. "A couple of them sneak glances in my direction, a couple do everything possible to avoid looking my way, & a couple just stare in a bold way, seeming to try to get a read from me, to figure me out," he noted. "It was painfully obvious who wanted to be on the jury, w/ catered answers to placate both the defenses queries & those of the prosecution. . . . You could see, by the way that they looked at me, that they couldn't wait to vote yes for the DP [Death Penalty]."

Most hurtful, Longo added, was listening to those who did not want to serve on his jury. These were the people, all of them dismissed by Judge Huckleberry, who announced that they were certain Longo had killed his whole family, that nothing could convince them otherwise, and that he therefore deserved to die, the sooner the better.

This was the first time since Longo's arrest in Cancún, fourteen months earlier, that he'd encountered anyone outside the criminal-justice system (myself excepted), and he was shocked to hear average citizens—he described them as "school teachers & dental hygenists & bank loan officers"—speak their minds. "The comments," he wrote, "seered through me & hit my heart pretty hard." He seemed incredulous to discover that people hated him. "I don't believe that I am this horrible person, this pariah," he wrote. "I feel like I still have something to offer."

In the end, Longo's jury consisted of eight women and four men, including a nineteen-year-old who worked as a deejay and a sixty-two-year-old retiree with a silver crew cut. It was a jury, Longo said, that seemed hungry to kill him. But he insisted that he had no intention of pandering to the jurors with maudlin displays of contrition or by begging for mercy. That was beneath him. "I will live or die by the truth," he declared. "Nothing matters more right now, in my life, than that. If it kills me to tell the truth in every matter, so be it, I'm a dead man."

THIRTY-THREE

THE OPENING STATEMENT for the prosecution was delivered by
Steven Briggs, one of the two lawyers representing the state. No
one in Lincoln County had much experience prosecuting a capital
case, so Briggs had been imported from the Oregon attorney general's office in Portland. He was thirty-nine years old, with a courtroom reputation for even-tempered aggressiveness and meticulous
preparation. He had a narrow face, an aquiline nose, and an air
about him of tightly coiled intensity, as if he were set to run a hun-
dred-meter dash. When speaking over the telephone with Longo,
my code name for Briggs was "the Greyhound."

His co-council was Paulette Sanders, forty-three, the chief
deputy district attorney of Lincoln County. In court, Sanders
appeared nearly as focused and ardent as Briggs, her dusty-blond
hair usually pulled into a no-nonsense ponytail, her few glances
toward Longo disgust-filled and withering. Both prosecutors were
unapologetically seeking the death penalty. "I've never seen a
harder push for death," Ken Hadley once commented to me during
a courtroom recess. "It makes me sick. They're practically drooling
for it."

Briggs spoke for half an hour. His tone was restrained and matter-
of-fact, though with a honed enunciation—the last syllable of every
fifth or sixth word snapped short—that left a distinct aftertaste of

fury. "December 16, 2001, was a cold night," he began. He displayed to the jury two framed photographs, one of MaryJane, another of the three Longo children. "This is what they looked like before that night," he said.

Longo had worked the late shift at Fred Meyer, Briggs explained, and when he arrived at the condominium he'd rented along the Newport bayfront, his family was asleep. It was about 11 P.M. "The defendant went over to his refrigerator," said Briggs, "poured himself a glass of wine, took a piece of New York–style sharp cheddar cheese, drank his wine and ate his cheese, and then he went to his wife for the last time. He placed his hand on her throat and he began to squeeze. In a violent struggle, she fought for her life."

The struggle, said Briggs, lasted three or four minutes, during which Longo bruised his wife's neck, bloodied her nose, and stripped off her nightgown. He then strangled all three of his children—two-year-old Madison, three-year-old Sadie, and four-year-old Zachery.

"When he was finished," Briggs continued, "the defendant went back to his wife, MaryJane, grabbed her naked, lifeless body, and dragged it over to a suitcase, where he pushed it into the suitcase and folded it up and squeezed it down and zipped it closed. He then reached down and picked up his two-year-old daughter, Madison, placed her in another suitcase with some clothing and diving weights, and he zipped that closed. He then picked up those suitcases and took them down to the water. Standing on the dock, he tossed the body of his wife and youngest child into Yaquina Bay, where they sank to the bottom."

Afterward, Briggs said, he loaded the bodies of his two older children into the family minivan and drove fifteen minutes south. He placed a large rock inside a pillowcase and tied the case to Zachery's leg. He did the same to Sadie. He wrapped them together in a black comforter and tossed them off the Lint Slough Bridge. At four-thirty in the morning, Briggs added, a man named Dick Hoch exchanged a few words with Longo while he was parked on the bridge. Two days later, Zachery's body bobbed to the surface. It was another three days before police divers found Sadie at the bottom of the lake.

According to Briggs, Longo had been planning this crime for a long time. He'd had an extramarital affair a year and a half earlier, during which he told his wife that she wasn't fun anymore, and that he no longer loved her. Six months before the murders, while his family was living in a warehouse in Ohio, Longo had downloaded from the internet a sixty-page booklet called *Hit Man On-Line*—"an instruction book on murder," according to the first sentence of its preface.

Longo also seemed to be preparing to alter his identity and possibly leave the country. Among his possessions left behind in the Ohio warehouse were a book titled *The Modern Identity Changer* and a Spanish phrase book. A few weeks before the murders, while living in Oregon, Longo printed from various internet news sites the obituaries of four men, all of whom were about his age. On some of the obituaries, Longo had written the deceased's Social Security number. He also lifted a credit-card receipt from the cash register at Starbucks—the receipt he later used to purchase a plane ticket to Mexico.

After killing his family, Briggs said, Longo stole a car, exercised at the gym, played volleyball, showed up for work, and rented movies. He tossed five garbage bags filled with his family's belongings into a dumpster. He even attended a Starbucks Christmas party, where he gave away a bottle of MaryJane's perfume and told several people that his wife had left him for another man and had taken their children with her. His coworkers felt so sorry for him that they wrapped up some pizza and sent it home with him.

Once Zachery's body appeared at the surface of Lint Slough, Longo fled Newport, drove to San Francisco, and later flew to Cancún. There, he impersonated a *New York Times* reporter, and ten days after killing his family, said Briggs, Longo was "drinking beer, dancing, swimming and snorkeling, and making love to a German girl in a cabana on the beach."

Nothing that Briggs said during the opening statement, Longo wrote, particularly concerned him. What worried him most was the presence of his family: Joe, Joy, and Dustin were all spectators in court. Joe sat stone-faced through Briggs's statement, his arm around Joy, who clutched a sheaf of religious booklets and quietly wept. "I really don't want them here," Longo wrote. "I know that I'm their son & they are hopeful that my life course will change, & they feel that support is necessitated, but at what point does support become pointless? I'm going to prison for the rest of my life."

Also in attendance were two of MaryJane's younger sisters, Sally Clark and Penny Dupuie, as well as her sister in law, Cathy Baker. They remained in Newport for almost all of the trial. One evening, I encountered the group in a local restaurant. They knew who I was—they'd worked closely with the prosecutors, who had evidently informed them of my connection with Longo. In court, they had occasionally fixed me with hostile stares.

At the restaurant, Dupuie told me that her hatred of Longo was so all-consuming that she was scarcely able to function. "I want him to feel uncomfortable," she said when I asked why she had come to the trial. "I want him to know we're here." Baker was less restrained. "I want to kill the motherfucker," she said, her eyes wide with anger. Then they informed me that because I'd been friendly with Longo, because I'd given him so much attention, I would soon burn in hell with him. All I could do was nod, meekly; there seemed no appropriate way to reply.

Longo knew precisely who was watching him. The courtroom's spectator section and defense table were only a few feet apart, close enough for observers in the first row or two of benches to occasionally make out the whispered comments between Longo and his attorneys. All Longo had to do was turn his head slightly, and he could see, through the corner of his eye, everyone in attendance.

By the start of the guilt phase of his trial, on March 10, 2003, it had been nearly two years since Longo had seen or spoken with his parents. Soon after they walked into court on the day of Briggs's opening statement, Longo glanced at his father but felt so ashamed, he wrote, that he quickly averted his gaze when Joe looked back. He did hold eye contact with his brother, but was stung by the "direct hatred" that appeared on Dustin's face—"I would have preferred for him to strike out physically," he wrote, "than to be so readable in his expression." As for his mom, the prospect of catching her eye was too daunting. "I couldn't go there," he wrote.

Longo and his parents had exchanged a few letters, all of which Longo eventually sent me—the original letters from his parents and photocopies of his own, which he'd had his lawyers make before mailing them. (Because Longo's parents were assisting the defense, he could swap mail with them via his attorneys, ensuring secure communication. Longo sent his letters to me through the standard jail-mail system, without inspection by Krasik or Hadley, much to their chagrin.) Longo's brother, during the year Chris had been in jail, had sent only a single message, a hundred and thirty words long. "I really don't know what to say," Dustin wrote. "I can't begin to imagine what you must be going through right now or what thoughts are running through your mind every day."

Longo's first contact with his mom and dad came two weeks after his arrest in Cancún, when he mailed them a one-page note. His tone was rueful and apologetic. "I do want you to know that none of this has anything to do w/ how you raised me," he wrote. "I think that you did better than most parents could ever imagine. I love you for that." He told them not to worry about him and "not to dwell on the whys or hows." He signed it: "Your Son, Chris."

In their terse reply, one handwritten page, his parents said that their contact with him would have to be limited because he was still disfellowshipped. "Hopefully some day we will understand what has happened," they wrote. "We did want to point out to you that if you are truly repentant you can take steps to mend your relationship with Jehovah. He is a merciful God." They signed it simply, "Mom & Dad."

"We have never stopped loving you," his parents wrote in their next message, which was two paragraphs long and printed from a computer. "Hopefully you realize that being honest and forthcoming with others and especially with those trying to help you is vitally important. . . . Our lives have been torn apart by what has happened."

Longo's response was long, nine pages, and petulant throughout. He sensed an undertone of presumed guilt in his parents' letters, and he felt offended, he wrote, "that you can even find me culpable of such charges." He said that he was already being completely honest and wasn't holding anything back. "I've learned that nothing is worth being dishonest about. I thought that I needed to protect my family, but I protected them in the wrong way and it ended up being no protection at all." His parents' expression of love, he added, felt rote, given purely out of a sense of familial duty. "It's possible for you to say you love me," he wrote, "but actually despise me."

"We do love you," his parents wrote back, "and not just out of obligation. One who loves only out of obligation doesn't hurt so much when their child is hurting or in trouble and believe me we hurt very much and all the time." They mentioned that they'd taken some of the items Longo had left in the Toledo warehouse and held a tag sale. "We've saved your suits," they added, "because we figured you would need at least one to wear to your trial."

"I do want to say that I'm very sorry for my last letter," Longo replied. "While it was a dumptruck load of what I was feeling, much of it wasn't fair."

"It sounds as if you have already done some soul searching," his parents answered. "Chris, there is something we hope you will really think about. You said the reason you had wanted to go to court was to set the record straight. From our vantage point it is pretty clear that your overpowering desire to have others think well of you is at the root of your problems." It was their last letter before the trial began. They signed it, "Love, Mom & Dad."

I'd met Longo's family at the rear entrance to the courthouse, a few hours before Briggs was scheduled to give his opening remarks. It was scarcely past dawn, and the doors were still locked. The limited number of spectator seats in Courtroom 300 were to be given away first come, first served, and the Longos—Joe and Joy, along with Dustin and his wife, Precious—wanted to ensure that they'd all have spots. (A few days later, Joe and Joy, whose only three grand-children had been murdered, would be granted victim status and given reserved seats. Dustin and Precious, however, had to wait outside with the rest of us.)

Joe had the husky physique of a former football player. His sil-ver hair was slicked back, not a strand out of place, and he had on a stylish black turtleneck. Joy wore a conservatively cut skirt and blouse, and plastic-framed reading glasses. They were both drink-ing McDonald's coffees and hunching their shoulders against the chill March air.

I introduced myself. Joe said he'd heard about me through his son's attorneys. He asked how I knew who they were, and I said that Dustin looked so much like Chris—the same boyish face and chiseled nose—that he'd given them away. Joe, in a valiant attempt at levity, said, "Well, we told him to wear the fake nose and glasses." It was the type of wit Chris had often employed—a little humor to sandpaper a potentially uncomfortable moment—and I laughed appreciatively and we shook hands.

Joe asked where I lived, and when I said I was from Bozeman,

Montana, Joy piped up and said that they were once in Bozeman. She even recalled the name of restaurant they'd eaten at: Frontier Pies. I asked how long ago that was, and she said, "Oh, Chris must have been fourteen." They were driving across America, she said. As she mentioned this, I recalled the letter in which Chris had described the family vacation. Joe had appointed him chief navigator, and Chris had read the map as they drove through the Dakotas and Wyoming and Montana, listening repeatedly to the *Out of Africa* sound track, talking with his father late into the night as the rest of the family slept.

"Chris told me about that trip," I said to Joy. "He mentioned that it was one of the highlights of his youth." I explained a little of how my relationship with Chris had started, and how it had grown, and how I found him to be charming and polite and likable.

I'd meant that last part as a compliment, but Joy looked at me strangely, her lips thin. "Yes," she said, "everybody likes Chris." She said this in a flat, horrible way—not acidly, not sarcastically—but with this emotionless tone that said to me: Be careful, he's not so benign as he appears; he's already crushed me, and he'll crush you too. And then her eyes glassed over, and I looked away and I was quiet. Joe was standing strong, but Joy, I saw, was a complete wreck.

"It must be hard," I eventually said.

"I don't know if I'll ever get over the shock," she said. And that was it. There was nothing more to add. We just stood there in the early-morning dampness, shuffling foot to foot, Joe and Joy sipping their McDonald's coffees, all of us waiting for the courthouse doors to open.

THIRTY-FOUR

AFTER BRIGGS COMPLETED his opening statement, it was the defense's turn. Krasik and Hadley, however, received permission to delay their opening statement—another unusual tactic. Neither man spoke about the crimes at all. Their plan, as Krasik expressed it to me after court that day, was to first allow the prosecution to present all their evidence. Only then would the defense attempt to challenge and rebut it. Longo's explanation for the murders, at least for now, would remain a mystery.

The state's case sped along quickly. The first witness was FBI agent Daniel Clegg, who confirmed, in a voice as crisp and confident as a newscaster's, that on the flight back to the United States after he had arrested Longo in Mexico, Longo had openly confessed to all the murders and explained that he'd sent his family to "a better place."

Denise Thompson, the babysitter who had identified Zachery and Sadie's bodies, also testified. She said that she ate lunch with Longo a few hours after Zachery's body had been found, although the discovery wasn't made public until much later that day. At this lunch, according to Thompson, Longo said that he had just taken MaryJane and the kids to the Portland airport. His wife had left him, he told Thompson, for "a guy who made lots more money." He even had a name for the man: Ron Gibson, a reporter for CNN.

(The network had no such reporter.) He said that MaryJane's affair with Gibson had been ongoing for years. The kids, Longo added, called him Uncle Ronnie, and he suspected that Madison might really be Gibson's child. Longo did not seem upset, Thompson said—"he was very calm and rational."

Thompson, on the other hand, felt terrible for Longo. She was concerned for his well-being, she testified, and invited him to Christmas dinner with her own family. Two days later, Thompson said, she was watching the TV news and saw a picture of the boy who'd been found in the lake. It looked a lot like Zachery, she thought. She sobbed on the stand when asked what it was like to look at the police photos and identify his body.

The day before Zachery's body was fished out of Lint Slough, two room cleaners working at the Newport Motor Inn discovered in the motel's dumpster several black trash bags filled with items that weren't typically thrown away, including photo albums, scrapbooks, stuffed animals, a woman's wallet, and brand-new clothing with the sales tags still attached. When the maids—both of whom testified—pulled the bags out of the dumpster and looked through the photos, they recognized the family as one that had stayed in the motel a few weeks earlier. They thought that perhaps the belongings had been mistakenly discarded, so they saved them. Later, when Longo became a murder suspect, the police confiscated the property.

These items were now brought into the courtroom. Everything was in Ziplock plastic bags labeled with red stickers that read STATE'S EVIDENCE. The photographs and the two scrapbooks, one detailing Madison's infancy, the other Sadie's, were shown to the jury. (Zachery's baby book was never found.) Each juror opened the bags, looked at the photos and the baby books, then passed them on. It took about an hour for all twelve jurors to have a turn. The courtroom was silent during this time, save for the flipping of scrapbook pages, the scratch of a juror taking notes, or the muted sobs coming from Joy Longo and MaryJane's sisters. By the end of the day, the area around the victims' seating section had become a midden of balled tissues.

Chris Longo sat in his chair with his nice suit and his neat hair and his wide-angled ears, the top of the left one oddly squared off, as if cut with a straight razor. He fiddled with one of Krasik's pens and affected a pose of unconcerned serenity. "I'd never felt more like I was in a fish bowl," he wrote about the hour of silence. He didn't know how to act. Or, rather, he was resolutely determined not to act. "I want this whole process to be honest & w/o additional drama or facades or put-ons," he wrote. The result, however, was that he appeared entirely impassive. This was not at all what he wanted to convey, but he felt uncomfortable, he wrote, expressing anything other than "the rigidity that I feel trapped in."

I usually sat directly behind Longo, in the second row of spectator seating (the first was reserved), and I spent many hours staring at the back of his head, observing every bob and nod while we both listened to the proceedings. We never spoke to each other in the courtroom, but each morning, as he was marched into court, and each afternoon, as he was escorted out, he'd glance over to me and briefly furrow his forehead and purse his lips to convey a look, he told me over the telephone, that was intended to say, "Can you believe this?" The expression I typically returned, as Longo described it, was that I smiled without smiling—I acknowledged him using only my eyes.

Longo occasionally called while the local news was on, and when he did, we'd watch the trial reports together—Longo viewing it on the TV in the jail's day room, which is also where the phone is located, and me watching it from my Newport rental house. Inevitably, the reports would contain video clips of Longo sitting in court, interspersed with still shots of his family, and Longo usually provided a running commentary as we watched.

"I don't know if I've got a good shave or not," he said, as his face appeared during a report on Briggs's opening statement.

"It's one of our wedding pictures," he noted, as a photo of MaryJane and him was shown. "It's not one of my favorites, that's for sure."

"You're right behind my dad, aren't you?" he said, as the camera panned across the courtroom spectator section.

"That's my balding forehead," I chimed in.

When a commercial came on, with an attractive woman demonstrating a hair-care product, Longo kept up his review. "I like this much better," he said.

The photographs and the baby books that the jury passed around were in all ways unexceptional. The snapshots were of moments almost any family could've captured. Zachery hiding inside a hollow tree stump. Sadie with chocolate on her face. Madison being held by Zachery. Chris riding a jet ski. MaryJane, pregnant, standing in a hayfield. Zachery and Sadie sitting together in a laundry basket. Chris and MaryJane playing Scrabble. Zachery blowing on a dandelion.

The prosecution seemed to be hoping that, through these photos, members of the jury would form a familiar connection with the Longos. Perhaps some jurors would see in them shades of their own lives. It was the ideal setup for the next group of photographs. These pictures were not passed around. Instead, they were projected onto a screen. The courtroom lights were dimmed, and then, one after another, enlarged to life-size, the police photos were displayed.

There was Zachery's body, moments after he'd been lifted from Lint Slough. He was lying on a grassy bank, his right leg bent beneath him. His skin was the color of plaster. Foam spilled thickly from his nostrils—typical, the jury was told, with a body that has spent time in water. There were reddish cuts around his lips and ears. According to the autopsy report, these were due to "marine life postmortem feeding activity."

One photo of Sadie was taken with an underwater camera. The

image was slightly fuzzy, and it took a while for the elements in the picture to sort themselves out. There was a leg, a skinny white leg, practically glowing against the gray waters. Something was tied to it. A sort of cloth, swirling with colors—a pillowcase, cinched tight, bulging with a hidden weight. Sadie's body neither rested on the bottom of the pond nor floated skyward. She was hovering in between; "neutrally buoyant," as Briggs described it.

Madison was shown inside her suitcase. The case had just been unzipped when the photo was snapped. Piled in the center of the suitcase was a mass of clothing, sixty items in all, including bathing suits, T-shirts, diaper covers, and socks. There was also a five-pound diving weight. Practically lost in a corner, curled into a semicircle, was the two-year-old's body. She appeared uninjured—marine life had yet to penetrate the suitcase and begin feeding—and seemed to be resting peacefully, as if settled into a nap.

Longo never looked at these photos. He made something of a show of this, averting his eyes, turning his head, wrinkling his nose as though repulsed by a smell. Krasik, he wrote, wanted him to look at them—"if for nothing else than to get a natural reaction from me that would undoubtedly display emotion." But he refused. "My own nightmares," he explained, "need no suplementing."

Indeed, in recent months he'd been having disturbing dreams. I asked him to write a few down, and he did, eventually recording fourteen of them. These were strange jottings, the handwriting messier than usual, as if he'd scrawled his impressions immediately upon waking.

In one dream, his family all fell over the side of a boat and began to sink. Longo dove into the water to attempt a rescue, but ran out of breath before he could reach them. In another, his family tumbled into a deep hole in the living-room floor. "I couldn't find the kids or MJ again," Longo wrote. He had a recurrent dream in which his tongue had grown so thick he could no longer talk. He witnessed a fatal car crash. He saw a boy drowning at the beach. "I went out & got him," Longo wrote, "but everyone came running, grabbing him from me, saying that I was trying to drown him like my family. They went to get the police & I ran."

The final photos the prosecution displayed in court were of MaryJane. Until this moment, the picture of MaryJane that I'd kept in my mind was one that her sister Sally had distributed to the media. The photograph had been taken by Zachery, from the rear seat of the Montana minivan. MaryJane, in the front passenger seat, had turned around to face her son. The grin on her face was wide and joyous. She was wearing fashionable sunglasses; her hair was windswept, blown about by the van's open windows.

Now, projected on the screen, was a picture of a green, soft-sided suitcase placed atop a medical table inside an autopsy room. The suitcase was open. In it was MaryJane's body. Her head was shoved into one corner; her body was bent, folded, flattened. She'd been transformed into a rectangle of flesh. Her eyes were hidden beneath a forearm. A foot had popped out and hung over the suit-case's side. "She was found in such a way," the medical examiner noted, "that she couldn't have put herself into that position."

The next picture was a close-up of her face. There were purple bruises on her neck and cheeks and just below her left ear. All of the bruises were the size of fingertips. There was a deep wound on the bridge of her nose. Around her eyes were dark, weblike contusions called petechial hemorrhages—a result of the force and ferocity of the strangulation, which had caused dozens of facial capillaries to burst as MaryJane struggled for air.

Both of these photos hung in the darkened courtroom for a few extra moments. What had been done to MaryJane, it was obvious, was unforgivable—as terrible as the kids' pictures had been, it was the graphic images of MaryJane that eliminated any notion I'd had that the killings were somehow motivated by love or compassion. MaryJane's murder was clearly a violent and frenzied act. And Longo had pleaded guilty to it.

After the photos were taken away and the courtroom lights switched on, I stared at Longo's hands, emerging pink and freckled from the sleeves of his suit. His fingers were thinner and longer than I'd remembered. His nails looked manicured. I could see a lattice of veins pumping blue beneath his skin. I visualized those hands encircling MaryJane's throat. I pictured them stuffing her into a suitcase. These images were so indelible, and so horrifying, that they seemed to have thrown a switch in my head. This was the beginning of the end of my friendship with Longo.

Briggs called each of the police divers to the witness stand, one after another, to describe what it was like to discover the bodies. Then the two suitcases were brought into the courtroom. Briggs pointed out that in MaryJane's there was still the outline of her body, formed by silt and sediment that had worked its way through the zipper. He displayed the rocks that had held Zachery and Sadie underwater. They were bowling-ball sized, black with yellow and white lichen. He showed the jury the comforter the two children were wrapped in before they were thrown off the bridge.

Dick Hoch arrived in court in blue jeans and a mechanic's shirt, with the sleeves rolled up and a pen in the pocket. He was clearly uncomfortable on the witness stand, and at times a bit grouchy, but in a way that lent veracity to his testimony—he wasn't grandstanding in the least, and seemed to want to say his piece as quickly as possible, then return to work. He explained that he was driving his pickup truck toward the coast at about four-thirty in the morning on December 17, 2001, on his way to clearing sand from his customers' driveways. Approaching the Lint Slough Bridge, he noticed a red minivan stopped atop it. Hoch pulled alongside and asked if he could help. A youngish man who resembled Longo was in the driver's seat. He said that everything was fine, so Hoch continued on to work. Days later, when he heard that two bodies were found beneath the bridge, he called the sheriff's office.

Linda and Lawrence Crabb, an elderly husband and wife who were staying in the condominium directly above the one occupied by the Longos, both testified that on the day the prosecution said the murders occurred, they heard loud noises in the middle of the night. The noises, which woke them both up, lasted for about ten minutes and came from either the condo below or the one next door. "It sounded like someone was moving the furniture around," Lawrence Crabb said. With this statement, the twelve jurors, rocking quietly in the office-style swivel chairs that crowded the jury box, all scribbled in their notepads.

To help establish the date of the crimes, Larry Hammons, the harbormaster of the marina adjacent to the condominiums, was brought into court. He testified that when he reported for work at 8 A.M. on December 17, he immediately noticed that a pipe running alongside the docks had been broken. Water was spraying all over—"like a fountain," Hammons said, waving his hands in circles. Ten days later, when the sheriff's office dive team searched the marina, the two suitcases were found in the water directly below the spot where the pipes had been damaged.

When the deputy state medical examiner, Dr. Cliff Nelson, was called to testify, he stood in the jury box with a wooden pointer. While photos were displayed on the nearby screen, Nelson indicated areas of note and discussed the particulars of the autopsies. There was green Play-Doh, the jury was informed, under both Zachery and Sadie's fingernails. Sadie's toenails were decorated with "pearlescent polish," the same polish MaryJane wore on her toes, as if the two of them had painted their nails together. Mary-Jane had fifty-five grams of partially digested food in her stomach, including popcorn. Lactation was evident in her breasts. She weighed one hundred and ten pounds.

Nelson explained that the children's bodies did not display any evidence of their having struggled against an attacker, indicating that the victims may have known and trusted the person who killed them. Except for the fact that Zachery was dead, he was found, according to Nelson, "in amazing condition." He exhibited no signs of mistreatment; there were no external injuries. His hair looked professionally styled. "He appeared to be perfect," said Nelson, and this comment seemed to strike everyone in the courtroom like a blow to the chest. Joy Longo pushed her way past the other spectators in her row and rushed out of the room. Chris clenched his jaw and trembled, and then two streams of tears flowed down his cheeks, unchecked.

When Judge Huckleberry asked the defense if they'd like to cross-examine the witness, Krasik apparently concluded that there was only one proper response. He said, "No questions, Your Honor."

THIRTY-FIVE

THAT WAS THE prosecution's case. They presented seven days' worth of testimony, then rested. All of the material was circumstantial, and there was no discussion of motive, but merely by reviewing Longo's actions after the crimes—lying, running—and offering a believable witness who placed Longo at the spot where Zachery and Sadie's bodies were found, the weight of the evidence, at least to me and all the spectators I spoke with, seemed damning, especially when added to the fact that Longo had already admitted to being a murderer, twice over.

Now it was the defense's chance. There wasn't much Hadley and Krasik could do. They called Oregon State Police detective Roy Brown to the stand, and Brown, who'd twice interrogated Longo following his arrest in Cancún, said that Longo "did not specifically" admit to killing Zachery or Sadie. They called Lincoln County Sheriff's Office detective Patricia Miller to testify, and she confirmed that during the hunt for Longo, vehicles with KIDVAN plates were spotted all over the United States—"like Elvis sightings," Krasik noted. She also verified that investigators had found no blood anywhere in the Longos' condominium.

Rebecca Cohen, a librarian at the Newport Public Library, testified that the Longo kids, escorted by MaryJane, had visited the library "several times a week" to read children's books. "They were never a problem," she said, holding a hand to her head as if in the grips of a migraine. Three other witnesses agreed that there was a rather loud party at the condominium complex on the night Linda and Lawrence Crabb heard all the odd noises.

Essentially, though, the defense had only one person to place on the stand. "We call Chris Longo," said Krasik, and Longo rose from the defense table and walked to the center of the room. He faced the clerk of court, Christine Bond, raised his right hand, and solemnly swore to tell the truth, the whole truth, and nothing but the truth (Bond was not an adherent of the "so help you God" addendum). Longo then proceeded to the witness box and spelled his name for the record.

While he did so, I looked at the spectators. Penny Dupuie covered her eyes, as if the sight of Longo's face was too much to bear. Joe Longo appeared outwardly calm, his fingers intertwined and his thumbs winding slowly around one another. Joy's hands were clasped as well, though pressed so tightly together I could make out the flicker of her pulse at the base of her wrists. Dustin's wife, Precious, was seated next to me, and she mumbled, "Oh God, oh God, oh God," under her breath.

Longo sat in the witness box with his forearms resting on the table in front of him—his hands, by command of the courtroom officers, had to be visible at all times, so that he could not furtively detach his Band-It. He answered Krasik's questions thoroughly and methodically, usually in complete, grammatically correct sentences, his voice a monotone, his gesticulations minimal. His eyes remained fixed on his lawyer. He never looked at the jury, not so much as a glance. About half the members of the jury—the women, it seemed, more than the men—stared at Longo unabashedly, as if at a circus sideshow. The others glanced around the courtroom, looking anywhere but at him.

Longo spoke a little about his childhood. He explained how his mom became a Jehovah's Witness. He described how he first met

MaryJane, and the way their courtship progressed, and why he moved out of his parents' house, and where he proposed marriage. After forty-five minutes of this, Briggs grew frustrated—to him, this all seemed irrelevant to the issue at hand—and he stood and issued an objection. "I think we've gone on long enough," Briggs said.

Huckleberry swiftly and firmly overruled him. "This is a capital case and every dispensation should be given," he said, and from then on, it was clear that Longo would be able to say whatever he wanted for as long as he pleased. Briggs, chastened, was reduced to scribbling notes on his legal pad, often frantically, holding two pens in his hand at the same time, a black and a red, and switching back and forth between colors.

Meanwhile, Krasik took full advantage of the judge's permissiveness. He leaned back in his chair, hands behind his head, and lobbed tough but friendly questions toward his client. He prodded Longo about his various lies and wrongdoings, better for the defense to do this, Krasik evidently figured, than the prosecution. It all seemed like an extended counseling session. "I want the jury to see him as much as possible, to get used to him," Krasik told me during the courtroom recess for lunch. "The longer they see him, the less likely they are to kill him."

And so Longo told, in elaborate detail and exceptional length—many of his phrases repeated, word for word, from his letters to me—of the camera-store theft he committed to pay for an engagement ring, of the birth of his children, and of his jobs with Publishers Circulation Fulfillment, Fireplace & Spa, and as founder of Final Touch.

Longo gradually became more relaxed on the stand, smiling a few times and occasionally chuckling, though he still didn't engage the jury or convey even a modicum of regret. He told me, over the phone, that he felt as though he were delivering his deathbed speech—"This is a man's dying words," he said—and that, appearances aside, he was "really stressed" about it and needed to swallow eight hundred milligrams of ibuprofen during the day to quiet his throbbing head.

The only time Longo revealed his nervousness was when, during a brief break in the proceedings, he returned to the defense table, picked up a metal pitcher, and tried to fill a cup with water. He dropped the pitcher midpour, spilling its contents across the table and floor. Krasik, aware of all the armed guards in the room, told Longo not to make any sudden moves. As Longo dabbed at the puddle with paper towels, he thanked the officers who were watching him for not triggering his leg-zapper. One of the side effects of a fifty-thousand-volt shock, he reminded them, is an involuntary emptying of the victim's bladder and bowels. That, he said, would be a real mess, and at this comment both Longo and the officers laughed heartily. MaryJane's sister Sally Clark, sitting in the spectator section, heard the laughter and promptly began to cry and dashed out of the room.

Over the course of hours upon hours of testimony, filling one day, then another, then a third, Longo related the story of how he stole the minivan, and of his affair with Jessica Meadows, and of the forged checks, the disfellowshipping, the demise of Final Touch, and the move to a warehouse in Toledo. Both Krasik and Hadley told me that never in their careers had they kept a witness on the stand for this duration. "It took Lewis and Clark less time to get to Oregon," one of the TV cameramen waiting outside the courthouse said during a recess.

Longo eventually commanded the courtroom as if performing a one-person play, with occasional cues tossed out by Krasik. His audience, spectators and jurors alike, seemed absorbed by the tale, though I noted that all but one juror—a woman with dark eyes and a piercing stare, sitting in the rear row of the jury box—soon stopped taking notes.

Longo talked about trying to sell the forklift, and how the police arrived, and the decision to try a midnight escape. Then he told of

driving away from Ohio—he and Zachery and the dog in the rented
moving truck, MaryJane and Sadie and Madison in the stolen minivan, the whole family on the road, their destination uncertain, hoping to leave behind their troubles and start a new life.

Most nights on their long drive west, Longo said from the witness
stand, the family camped out, paying a modest fee to pitch their tent
at a state park or a private campground. It was the best way to save
money. For meals, fast food was the norm—McDonald's, Taco
John's, Pizza Hut, Arby's. They crossed Indiana and Illinois and Wisconsin and Minnesota. By the time they were approaching South
Dakota, the family, according to Longo, was in good spirits. They
felt as though they were on vacation, Longo said, and all of their
stresses, with the exception of their financial situation, seemed to
dissipate.

The only other issue was with their family, Kyra. Longo realized
that the dog was going to be difficult to travel with, so one morning, early in the trip, he woke before his children and let the dog
loose on a nearby farm. When Zachery asked where Kyra had gone,
Longo told him that their dog, too, was having a holiday, only with
friends of her own, other animals. Zachery seemed to find this
explanation acceptable.

Their trip nearly ended at the South Dakota border. Longo
pulled the rental truck over at the state-line weigh station and the
trooper there asked for his driver's license. When Longo handed it
over, the officer ran it through his computer and, according to
Longo, immediately seemed suspicious. The trooper asked about
the items Longo was hauling, and where he was going.

At this moment, Longo testified, he felt "extreme paranoia." He
thought of all the outstanding arrest warrants that might bear his
name—for parole violations, for counterfeit checks, for the stolen
forklift and boat, for stealing the minivan. He was potentially facing
years of jail time. There was also the missing-persons report filed by

MaryJane's family, which he didn't know about. But nothing, apparently, was entered in South Dakota's system. The trooper let him go.

The weigh-station experience, and the fact that the truck guzzled fuel, led Longo to reconfigure his plans. He rented a storage unit in Sioux Falls under a made-up name, John Purty, and moved into it most of the truck's contents: furniture, clothing, a collection of framed animation cells. Anything that could be easily pawned—a DVD player, a TV, two vacuum cleaners, some scuba gear—he stuffed into the van or tied to the roof. The whole family was now together in one vehicle. Longo abandoned the moving truck in South Dakota, and Penske, the company that leased Longo the truck, soon reported it stolen.

The Longos visited Mount Rushmore and Devils Tower and Yellowstone National Park. They stopped at a prairie-dog town, and Sadie tried to name every dog that popped out of a hole, but eventually gave up and referred to them as "101 Dalmatians." Zachery marveled at the Old Faithful geyser—"The ground is spitting," he said—and asked his parents for permission to stand beneath it.

They ate fudge at a fudge factory and visited an Indian reservation and hiked in the mountains. They played a game in the car, to see who could come closest to guessing the population of the upcoming town. Longo kept precise tabs on their money, and was honest with MaryJane about how much they had left. They were both aware of how quickly their funds seeped away: $1,033 remaining on September 1; $639 on September 3; $492 on September 8.

On September 9, 2001, a week and a half after leaving Ohio, they arrived in Portland, Oregon, worn out from travel and in urgent need of income. Chris and MaryJane thought, for a moment, that the family would live here; they liked the urban feel and the proximity to mountains and beaches. But once they saw how high rents were and how meager the job market was for someone lacking a college diploma, they swiftly abandoned the idea. What Portland did offer, however, was a row of wholesale jewelry stores.

It was MaryJane, according to Longo, who first brought up the idea of selling her engagement ring. She did this a few days before they'd reached Portland, and Longo says he refused to even consider the notion. "That was the one thing I did not want to do," he testified. But as their money disappeared and MaryJane pushed harder, his resolve melted. So they stopped in a few jewelry stores in Portland. The three-quarter-carat diamond—"simple, yet precious," he noted, "like MJ"—was worth more than $3,000. Longo was hoping to sell it for at least half that. But the highest offer he received was $600.

His wife told him to take it. She said he'd just have to replace it later with a better one. Longo assured her he would. And so, while MaryJane and the kids waited in the van, Longo sold his wife's ring. He was paid in cash, in hundred-dollar bills. "This was the symbol that I had failed everything," Longo said.

The money did provide them with a bit of breathing room. They left Portland and drove to Seattle, but rents there were even higher and the job market more daunting, so they turned around and drove south. The best deal, they figured, would probably be an off-season vacation rental in a small town on the coast. They camped out on September 11, the day of the terrorist attacks in New York and Washington, D.C., and the next day rented a two-bedroom bungalow, fully furnished and cutely painted with lime-green trim, just a few hundred yards from the beach in Waldport, Oregon. "It was perfect," Longo said from the witness stand. For the first time in months, the family had a decent place they could call home.

It didn't last long. The rent on the Waldport cottage was $800 a month. Longo had arranged with the home-rental agency to pay by the week and had negotiated his way out of providing a deposit, but he still needed to earn money. He unpacked his computer and constructed a résumé. In the brief paragraph that mentioned Final Touch, he said he'd grown the business from "zero to over $1 million in sales in the first year" and that he had "a workforce of nearly 100." There was no mention of the company's demise. At the bottom of the résumé, in a section labeled INTERESTS, the first thing he listed was "Family."

The résumé was of little help. Longo attempted to find work with a photo shop. He tried at a fireplace distributor. He searched the internet, the newspaper, and the offerings at an employment agency. The only opening he could find was at the Starbucks inside the Fred Meyer department store. Longo interviewed for it, and was offered the job. It was part time and paid $7.40 an hour. He began work on Monday, September 24, 2001.

Longo thought of coffee-making as a low-status job, and he couldn't stand the fact that he was forced to do it. "It was driving me crazy," he said in court. "It was not fulfilling in the least." So he invented a new life for himself. He told his fellow Starbucks employees that his family was well off. They'd come to the Oregon coast to take a break from their hectic, big-city lives. The Starbucks job, he explained, was just a way for him to kick back and kill a little time. Also, he liked the coffee.

Longo declined, though, to wear the standard Starbucks uniform of a short-sleeve polo shirt and khaki pants, and instead usually overdressed in slacks, a business shirt, and a tie. He wore a pager, and told the other employees he used it to keep tabs on his stock options. He talked about ski trips and scuba-diving vacations and his advanced knowledge of wines. "I wanted to give the impression that the job wasn't necessary," he testified. "I talked about things like owning a website, being an internet mogul."

He claimed that he earned $15,000 a month from a business called Zooweb, an online service that provided ratings and tips for most of the nation's zoos. (There is such a site; it's just not Longo's.) "I've got over 40,000 people using zooweb as their email address everyday . . . the money keeps flowing in," Longo wrote in an e-mail to Denise Thompson, a coworker with whom he'd become friendly.

It's nearly impossible to feed and shelter a family of five on $7.40 an hour. Longo pawned whatever he could, including his digital camera, his binoculars, and his wetsuit, but this netted him less than $400. Desperate, he took two crab traps that were used as decorations in their rental home and pawned them as well, receiving $10 each. MaryJane wasn't working; she needed to take care of the children. The Longos simply could not afford the rent, and a month after moving to Waldport they abruptly departed, leaving the last week unpaid.

The Longos relocated in Newport, close to the Fred Meyer, and stayed in a series of inexpensive motels, eventually settling into the Newport Motor Inn—the five of them living in a $20-a-night room. Their kitchen consisted of a microwave and a dorm-room fridge. They remained there for most of November. While Longo worked at the Fred Meyer Starbucks, MaryJane and the kids spent a lot of time at the McDonald's playground and the public library. They had less than $5 a day to spend on food. Mostly, Longo said, they ate ramen noodles and bread.

Krasik asked him in court why he didn't apply for welfare. "Public assistance is something that I would never go on," Longo answered. "I would literally steal before I went on public assistance." MaryJane, according to Longo, was able to withstand the food situation and the living conditions. He said that the children were also fine, though it's impossible to know if this was true. Elisabeth Young-Bruehl, one of the psychologists who studied Longo's writings, addressed this issue in her analysis of the letters. "He studiously presents them [the children] as happy and playful," Young-Bruehl noted. "But how could they be? Being dragged all over the place, living and sleeping in strange rooms, left hungry, etc."

What really upset MaryJane, Longo testified, was that they'd done nothing to revive their spiritual lives. There was a Kingdom Hall in Newport, but they did not attend services. Longo's excuse was that all their decent clothing was stuck in the storage unit in South Dakota. He didn't want any Witnesses to think he couldn't afford to properly dress his family. Also in the storage unit was most of their winter clothing; by early November it had grown cold on the Oregon coast, and the kids scarcely had enough to wear outside.

Longo solved these problems by cashing in some of the frequent-flier miles he'd amassed during his Publishers Circulation Fulfill-ment days. He flew to Sioux Falls, removed from the storage unit all the clothing that would fit into his bags, and flew back the next day. Before leaving, he mailed two greeting cards written by MaryJane—one to her sister Sally, and one to her mom.

The card Sally received was shown to the jury. The postmark said November 5, 2001. By this date, it had been ten weeks since MaryJane had contacted her family. Chris had been disfellow-shipped and was expected to remain distant from his mom and dad, but MaryJane, whose cell phone had been disconnected, knew that her family would worry about her and the children—and indeed, they'd already filed a missing-persons report. MaryJane may also have suspected that her sisters would visit her if they could, and she likely didn't want anyone to know she was living in a dingy motel room. The extent to which Longo coerced his wife into writing the cards is unclear, but it seemed that MaryJane agreed, at least on some level, to present a cheerful front and dis-guise her family's location.

On the cover of Sally's card was a picture of a teddy bear. The message inside was remarkably vague. "Hi!" it began, in MaryJane's looping cursive. "Sorry I waited so long to write but time goes by <u>so</u> fast. As I'm sure you guessed we moved. Chris sent out his resume and got a couple good bites so we went to check out the areas and decide where we wanted to move to. I still don't have an address or number to give you because he's in 8 weeks of training and then he could be sent to anywhere in the USA. We miss you guys. I hope all is going good with you."

MaryJane's family told the Michigan police about the cards, hoping it might assist them in tracking down the Longos. Instead, when the police read the cards, they concluded that MaryJane was voluntarily avoiding contact. The Longos, they determined, had simply moved away, as MaryJane had written. The police deleted them from the computer system. They were no longer missing persons.

In late November of 2001, Longo said, continuing his testimony, he was promoted to the home-furnishings section of Fred Meyer. He received a modest raise and full-time hours, and immediately began searching for better housing. Along Newport's bay-front walkway—probably the most valuable stretch of real estate in town—he found an upscale condominium complex called The Landing, whose manager, James Calhoun, was willing to work out a rental deal.

Longo told Calhoun that he was employed as a subcontractor for Qwest Communications, researching the demand for high-speed internet service in Newport. He said that he'd be living by himself, though his family might come for an occasional visit. "I wanted to give the impression of being a businessman," Longo testified. He informed Calhoun that Qwest would soon be sending him a very large check, with which he'd pay the bill. Calhoun accepted his story and didn't ask for a down payment or a credit-card imprint. Longo declined the maid-service option, which made the rent on his unit, number 211, $1,200 a month. "I was proud of the fact that I was able to get a place like that," he said from the witness stand. The family moved in on November 30.

As Longo was solving his housing issues, he was also working on his career prospects. He was intent, he said, on finding a corporate position, perhaps with Starbucks, but first he needed to be able to pass a background check. He was wanted by the police in both Ohio and Michigan. For a low-ranking job like the one he held at the Fred Meyer, his background wasn't inspected, but to secure a position that would allow his family to live comfortably, there was no way he could use his real name or Social Security number. And he couldn't invent new ones; that wouldn't pass a check either. Back in Toledo, he'd purchased a book called *The Modern Identity Changer,* and he decided it was time to apply the information he'd learned from it.

First, he used the public library to scan the obituary sections of various newspapers, looking for people who were born about the same time he was. He eventually found four. Next, he checked to see if each person had been registered with the Social Security death index, which is posted on the web. Some families, in their grief, neglect to do this, especially when the victim is young and dies unexpectedly. This leaves a viable Social Security number attached to a deceased person.

Three of the men who'd died had their names and numbers listed on the death index, but one did not: Alan Rae Swander of Albany, Oregon, who was born on April 20, 1974, three months after Longo, and died in an automobile accident on May 22, 2001. Swander's number was likely a good one to use—as far as the federal government was concerned, Swander was still alive—but Longo did not yet know the specific digits.

If someone is not on the death index, the quickest way to find his Social Security number is to access his death certificate. Proof of kinship, though, is often required to see this document. Another method is to hire an information vendor—scores of them are advertised on the internet—who will provide almost anybody's Social Security number for a modest fee, no questions asked, all major credit cards accepted.

But Longo didn't have a valid credit card; all the ones he owned were overdrawn. So a few days before he left Starbucks to work in the home-furnishings department, he stole a credit-card receipt. He said in court that he battled his conscience before doing this, but rationalized that it was "a small sacrifice for the greater good." He jotted down the card owner's name and account number, stashed the information in his wallet, and threw away the receipt.

The prospect of a more stable future, Longo testified, is the reason he lifted the credit-card number and had among his possessions obituaries with Social Security numbers written on them. He didn't remain in Newport long enough to become Alan Rae Swander, but he did use the credit-card data to purchase a plane ticket to Mexico.

As for the *Hit Man On-Line* booklet that Briggs had mentioned in his opening statement as proof of advance planning of the murders, Longo explained that it was merely one of hundreds of files he downloaded one night while the family was living in the Toledo warehouse. "It was kind of a last-ditch effort to make some money," he said. He gathered from the internet all the strange and unconventional information he could find, then created a home page and charged $12 a person to access it. The site was called NoToKnow.com, and Longo advertised it as "the best place for all of the forbidden, not to know, information on the web." In total, he earned under a hundred dollars. He could have made a fortune, he said, but he had to abandon the warehouse and disconnect his internet service.

For the last three weeks of their lives, Longo's family lived in relative luxury. The condominium in The Landing had a full kitchen, a washer and dryer, space for everyone to sleep, and a grand view of Yaquina Bay and the Pacific Ocean. "It was everything we needed," Longo said. "It seemed like a big answer."

They even made friends. Longo's coworker Denise Thompson and her husband, Macon, had two children about the same ages as Zachery and Sadie. The Longos invited the Thompsons to dinner at the condo. The Thompsons brought over a salmon, and the two families, Longo said, got along wonderfully. Macon and Chris discussed starting a business together, selling internet service. Denise and MaryJane later spent a few afternoons with each other. As an early Christmas gift, Denise knitted MaryJane a scarf. "It was great to finally feel normal again and have people over and let the kids play with somebody else besides each other," Longo said.

The condominium, though, was expensive. Even with Longo's higher salary and full-time hours, his monthly income, after taxes, was scarcely sufficient to cover the rent. Longo's paycheck, during the first week that his family lived in the condo, was devoted entirely to food, diapers, shampoo, dishwasher detergent, and other items related to the move-in. Longo told MaryJane that he'd made arrangements to stay in the condo long-term, paying the rent bit by bit from his salary. He explained to the manager of The Landing that his payment from Qwest had been delayed.

But there were little things the second week, too—the minivan needed gas; the kids hadn't been given a new toy in months; it was time for another round of groceries and diapers. Before the week was out, Longo's paycheck was gone again, with nothing left over for rent. He even spent the change in the minivan's ashtray. Once more, Longo went to The Landing's manager, James Calhoun, and spun a story. He was given a reprieve of a couple of days, but it was obvious that Calhoun's patience had worn thin.

Longo received another paycheck from Fred Meyer on Friday, December 14. It was for $170. Longo wanted to give all of it to Calhoun, but again the family needed some things, and Longo had also promised MaryJane that they would go out on a date. It had been more than a year since just the two of them had gone to dinner, and they'd made firm plans for Saturday night. They had even asked Denise Thompson to come to the condo and babysit.

On Saturday morning, the whole family went shopping. They went to the Fred Meyer store, where Longo had an employee discount, and bought milk, sugar, cheese, microwave popcorn, a roasting pan, baby wipes, a children's book, and a Lego set. A brief videotape, recorded by a Fred Meyer surveillance camera, was played in court, showing Chris, MaryJane, and the three children all together, pushing a shopping cart toward the store's exit.

Denise Thompson arrived at 6:30 P.M., and MaryJane and Chris left. MaryJane was wearing the scarf that Thompson had knitted her. The kids ate macaroni casserole and watched a movie, *The Little Vampire*. Zachery and Sadie were happy, Thompson later said, but Madison was upset because MaryJane wasn't there.

For their dinner date, Chris and MaryJane went to Rogue Ales, the local brewpub. They talked. MaryJane told Chris that it had been rough living in the motel room, but now that they were in the condo she felt better, more settled. She told him, according to Longo's testimony, that she was proud of him for earning a promotion so quickly at Fred Meyer.

But MaryJane also expressed a touch of suspicion. Even though Longo wasn't earning that much more money, they'd gone from a low-budget motel along the highway to a high-end condominium on the bay. It seemed too good to be true. She reminded Chris of his promise of complete openness and honesty, and asked him, bluntly, if there was anything he needed to reveal. She wouldn't be upset, she said; she just had to know.

Chris claimed, in his testimony, that he wanted to tell her everything. He wanted to tell her about the panic that was roiling inside him; that they were a day or two from being kicked out of The Landing; that their one valuable possession, the minivan, wasn't even legally theirs. But how could he admit this? MaryJane had just said she was proud of him. She said she was feeling better. So he told her that everything was fine.

"This is not a dream, not a facade," he said to his wife. Rather, he told her, this was only the beginning. They were back on track and everything was going to progress from here. He'd earn more promotions, he promised her. They'd return with full commitment to the Kingdom Hall. And their lives would once again be free of worries.

"We left dinner happy, arm in arm," Longo said on the stand. He spoke this part of his story while staring passively at the floor, as if studying the courtroom's carpeting, which was seawater green, flecked with strands of yellow and pink. Judge Huckleberry seemed to have followed his eyes, and was gazing at a similar spot. Briggs, at the prosecution's table, was still writing on his pad.

MaryJane and Chris saw the eight o'clock showing of *Ocean's Eleven* at the Newport Regal Cinemas. They were home, Longo testified, by 10:30. Zachery and Sadie were still awake, and Longo watched cartoons with them for a little while, until they fell asleep.

On Sunday, December 16, Longo slept late and then worked at the Fred Meyer from 2 P.M. until 11. At work, he thought about MaryJane's question from the evening before, asking if he'd been hiding anything from her. He didn't know whether to confess all he'd done or to try and maintain a delicate balance—admitting some things, hiding others, keeping her satisfied one day at a time.

By the time he drove home that night, he was severely depressed. His final deadline for paying the rent, Monday morning, was a few hours away, and he didn't have a dollar to spare. They would have to leave. But where would they go? Back to the fleabag motel, back to ramen and bread? And then what—more counterfeit checks, more arrest warrants, more running away? He understood, for the first time, that things would never get better.

When he entered the condo, his family was asleep. MaryJane was in the bedroom, with Madison on a comforter on the floor. Zachery and Sadie were on the fold-out sofa in the living room. The television had been left on, and he turned it off. He was hungry. He poured himself a glass of pinot grigio and ate a hunk of cheddar cheese.

Then he walked out onto the balcony. It was nearing midnight, cold and drizzly. Boats rested in the harbor, lines clanging against masts. The lights of the Yaquina Bay Bridge formed a bow against the sky. Longo broke down. "I remember looking out on the perfect setting, and knew we'd have to move," he said. "It was just set up to be a horrible week."

He wandered back inside. He brushed his teeth. He went into the bedroom and lay down next to MaryJane. She stirred and said, "How was your night?" and he said, "Fine," and kissed her good night, and she fell back to sleep. But Longo was wide awake. Thoughts of his family churned in his head. "I was thinking that they were in that situation too long with me," he said. "That they deserved much better. I didn't know if I could give it to them."

THIRTY-SIX

A FEW MINUTES before Longo returned to the witness stand to relate the final chapter of his story, Ken Hadley and Steve Krasik spoke privately with me. I'd been waiting outside the courtroom during a recess, drinking a soda from the RC Cola machine, when the two lawyers motioned for me to follow them. Hadley opened a wooden door with a blue sign that read COUNSEL / CLIENT CONFERENCE, and we squeezed into a windowless room. Krasik and I sat at the two-person table while Hadley remained standing.

Over the course of the trial, I had developed a good relationship with both men. I'd once accepted an invitation to a crab dinner at Hadley's house, and Krasik and I had met a couple of times at the local sushi restaurant to exchange trial gossip and share a meal. Now, in the conference room, the two attorneys appeared ill.

"We're about to cross the Rubicon," Krasik said. He'd removed his eyeglasses and was massaging his temples.

Hadley was blunter. "Chris is a dead man," he said, "if he tells this story."

The two attorneys, apparently stymied by Longo and conflicted over how to proceed with his defense, were attempting a different approach. They were asking if I had any ideas.

"We don't want to look like fools," Krasik said.

"We're just punching bags," Hadley added.

They asked if Longo had already told me the remainder of his tale. I shook my head no. Both lawyers implied that they feared they were blindly leading their own client to his death. This, I figured, may be why they hadn't given an opening statement—they did not want to lock themselves into a specific story. They asked if I had any notions, any at all, of how they might prompt Longo to change his mind, to confess to everything, to plead for leniency from the court.

"You're his friend," Krasik prodded.

"This would all be so much easier," Hadley said, "if I didn't like Chris."

"It would be such a waste to kill this man," Krasik concluded.

Then they fell silent. The air in the conference room seemed to drain away. I felt, at that moment, as though Longo's life was in my hands—that if I said the right thing, he'd be spared the death penalty.

I knew that a death sentence, even if the inmate lives out his natural life, is vastly different than a life sentence. In Oregon, death-row inmates live for years, as they pursue their appeals, in so-called supermax units, kept alone, locked down nearly the entire day, so sealed off from the world that not even sunlight is allowed to shine into their cells. All other inmates, no matter their crimes, can qualify for full-time jobs and a spot on a softball or basketball team, and can spend twelve or more hours a day away from a cell. "The difference between death row and the general inmate population," Hadley had said to me over dinner at his house, "equals the difference between the general inmate population and freedom."

I searched my mind for something to tell the lawyers. I had spoken to Longo for all those hours and read all those letters, but I still hadn't gained any true understanding of him. During the trial—especially after the autopsy photos had been shown—I'd found that the more I learned about Longo, the less I could state about him with any certainty. I now wanted to back off from our relationship;

a little distance, I thought, might help me arrive at some insight. The last few times I'd seen INMATE PHONE on my rental house's caller ID, I had been tempted to leave it unanswered. Mostly out of habit, I ended up taking the calls, though our conversations ended well before the time expired. One evening, over the phone, I'd even referred to our relationship in the past tense, as if my image of him had shifted entirely. "I thought you had some very good parts to you," I'd said. "I liked you." Longo's only response was a muted "Okay."

I had asked Longo a dozen times since his plea hearing if he was being entirely honest, and he'd answered a dozen times that he was going to live or die by the truth. There seemed no way to counter that. Now, sitting in the conference room with Hadley and Krasik, I sensed that I was at least partially responsible for what might occur. I wanted to help Longo, but I didn't know how. I felt a sharp, deep fear, as if I'd hit a patch of ice while driving and was sliding off the road—my mind was racing, but I was essentially helpless. All I could think of was running into the courtroom and yelling, "Stop!" What else could I do? Longo's course, I felt, had already been set. I wasn't holding anything back; I just had nothing to say. I nibbled on my bottom lip, shook my head, and apologized to the lawyers. Then we left the conference room and returned to court.

Longo continued his testimony. He was lying in bed, he said, next to his wife. It was past midnight. MaryJane eventually got up to use the bathroom, and as she returned, she noticed he was awake. She asked him what was wrong. Longo hesitated, and she asked him again. He said that he was thinking of the discussion they'd had on their date, when she asked him if he'd been hiding anything from her.

He said that he hadn't been fully honest with her. There were, he admitted, a few things he'd held back. She said she wanted to know. He began by telling her about the condominium—that he hadn't paid the rent, that he'd been lying to the manager, that they were going to have to move out.

She was upset, but wanted him to continue. He told her he'd recently stolen a credit-card number while working at Starbucks, to help establish a new identity. He told her that the Penske moving truck he'd rented in Ohio had been authorized only for in-state use. He told her that he'd taken two crab traps from the Waldport house and pawned them. At first, Longo testified, the conversation was fairly restrained. With each admission, MaryJane would react angrily, then quickly settle down and demand further information.

He told her that, in Toledo, he'd lied to the landlord in order to rent the warehouse, and that he hadn't really paid a year in advance, and that the building wasn't zoned for residential occupation. He told her that while they were living in Ohio, he'd once driven off after a fill-up, without paying for the gas. He told her that the last two credit cards they'd owned had been illegally acquired by forging his father's signature.

As the revelations mounted, MaryJane became increasingly anguished. She began to cry. "She just said quietly that she could no longer trust me," Longo testified. This was the first time she'd ever said such a thing—after the camera-shop theft, after the counterfeit-check conviction, even after the affair with Jessica Meadows, she'd always affirmed her belief in Chris's underlying integrity.

He told her more things. He couldn't stop; it was as if he'd reached a threshold and needed to absolve himself by releasing all he held inside. He told her the actual amount they'd received for the sale of their home. He told her how he had purchased the forklifts from a shady acquaintance, and that he'd suspected all along they were stolen. MaryJane asked about the boat, and he confessed that it, too, had likely been stolen. Then he told her that, while they were living in Toledo, he'd cashed another round of counterfeit checks.

With this, the confrontation turned physical. "She slapped me at one point, which she had never even come close to doing,"

Longo testified. He spoke this line the same way he'd delivered all the others—calmly, his eyes untroubled, as if he were relating an anecdote about someone else. But in the spectator section, at least, I felt a subtle change. We all seemed to shift our sitting positions at once, as though bracing ourselves for an impact, and I could hear the clunk of a courtroom officer's gun as his holster knocked against the wooden bench.

Longo kept on. "We'd never had an all-out argument," he said. "We'd never raised voices with each other. But she was raising her voice at me now." He said that he did not strike back. Madison, lying on a comforter on the bedroom floor, began to cry. MaryJane picked her up and held her.

She asked more questions. What about the jet skis? Longo told her that they, too, were stolen goods—that he hadn't won them in a raffle. What about the affair with Meadows? Longo insisted he'd been fully honest about that. What about other affairs? Longo said that one night, years before, when he was traveling for Publishers Circulation Fulfillment, he'd shared a romantic dinner with a woman in a Chicago hotel; afterward, when they had arrived at his room, he considered taking it further, but had decided not to.

MaryJane, according to Longo, was "borderline irate," but not yet finished. She demanded that he tell her everything. If he made a full confession, she implied, perhaps their relationship could be salvaged. "She actually started to act as though we could get to a point of reparation, that we could get beyond this," he testified.

So Longo told her the final thing: He said that he'd stolen the minivan. "That was it," Longo testified. "She lost any self-control that she had at that point. She just started crying heavily and told me to get out. I didn't want to get out at that point. I wanted to console her and I reached over to console and she went to hit me again." The look on her face, Longo said, was daunting. "I had never seen her look like that, like I was somebody completely different than she ever thought I was."

Longo then left the bedroom and went into the living room, where Zachery and Sadie were sleeping on the sofa bed. He lay down with them. He could hear the sea lions honking in the bay, he said, so he knew that dawn was approaching.

It was early in the morning on Monday, December 17, 2001. Probably it wasn't too much past 4:30 A.M.—the same day and time that Dick Hoch said he spotted what looked like a maroon Pontiac Montana minivan stopped on the Lint Slough Bridge.

"Where were you," asked Steve Krasik, "at 4:30 on Monday the 17th?"

"I was in the apartment," Longo answered.

"Weren't driving the Montana on a bridge?"

"No."

"Did you ever park the Montana on the bridge?"

"No."

He also insisted that he did not disturb any pipes in the marina around that time, despite the harbormaster's testimony that someone or something had broken them that very morning.

What he did do, he said, was sleep with his children. He wasn't scheduled to work that day until two o'clock. He woke around nine, when Zachery leaped onto his chest. His son wanted to know what he was doing in bed. "I just told him that this was my night to sleep out here," he said. "Madison was going to sleep with Mommy tonight and I was going to sleep with them."

He prepared breakfast for Zachery and Sadie, then he went to the bedroom to check on MaryJane. She'd locked the door, but it was an easy lock to pick—it could be done with a straightened paper clip—and after knocking a few times Longo let himself in. Both MaryJane and Madison appeared to be asleep. There was a foul smell in the room; MaryJane had vomited on the bed. He gently touched his wife, to see if she was okay. "She winced away from me pretty violently," Longo testified. "She was awake evidently when I walked in and she didn't want me touching her. She didn't want anything to do with me."

Longo carried Madison into the living room to sit with the other children and watch cartoons. She didn't want to eat cereal, Longo recalled—she was a fussy eater—so he gave her a piece of left-over pizza from the refrigerator. Then he returned to the bedroom to clean up the vomit, and MaryJane immediately went into the bathroom. He talked to her through the door. He told her that he wasn't going to work. He was staying home, he said, to make sure everything was okay, so they could talk and maybe come to a reasonable solution.

"She yelled through the door that I most certainly was going to work, that she didn't want to see me," Longo testified. He spent the remaining hours before work playing with the kids. They built a house out of Legos and created sculptures with green Play-Doh. MaryJane, he said, remained out of sight.

When it was time to leave for work, he told MaryJane, who was back in the bedroom, that he was departing. She said that she wanted to take the children out. She didn't say where, but Longo supposed it was to the library. She said she needed the car, which surprised him, as she was now aware it was stolen property. Longo didn't try to dissuade her, though; he wanted to be as accommodating as possible, to do whatever she wished to calm her down and attempt to mend their rift.

The whole family piled into the minivan, and MaryJane drove Longo to the Fred Meyer. She didn't speak a word to him during the ten-minute trip. She stopped in the parking lot, and Longo got out of the front seat and opened the minivan's sliding door and kissed each of his children. "Madison reached up with her Scooby toy to give me a kiss good-bye and I left them in the parking lot and went to work," he said. According to Fred Meyer's records, he punched in at 1:56 P.M.

It was a normal day at work, Longo testified. At five o'clock he had a dinner break. He called the condominium from a pay phone, but nobody answered. He worked until eleven and punched out.

MaryJane knew his schedule, and when he exited the store she was already there, waiting to pick him up.

Rather than driving to the front of the store, though, she'd parked in the middle of the empty lot. Longo walked to the minivan. Mary-Jane had already shifted over to the passenger seat; she was crunched up against the door, as if trying to stay as far from Longo as possible. Her legs were curled beneath her. She was wearing a white terrycloth robe but no other clothing he could see, not even shoes, though a pair of her hiking boots were in the footwell. "Everything was kind of out of the ordinary," Longo said. When he greeted her, she did not respond. He settled in behind the wheel and began to drive.

Normally, when MaryJane would pick up Longo, no matter how late, she'd bring the children with her. There was no alternative but to leave them alone, which she would not do. Often, they'd all be asleep in the back of the van, so the fact that Longo did not hear his kids was of no concern. It was only when he glanced in the rearview mirror and noticed that all of the car seats were empty that he became mildly puzzled. He assumed, he said, that MaryJane finally wanted to discuss things with him and had asked Denise Thompson to look after the children.

Longo attempted to initiate a conversation. He asked if Denise was babysitting; he asked if she wanted to drive somewhere and talk. She said nothing. "I was actually starting to get a little bit irritated," Longo said. He continued driving, back to the condominium complex and into the basement parking garage. He tried talking to her one more time, but MaryJane remained silent. Longo stepped out of the car and walked to the garage's elevator. MaryJane stayed in the passenger seat.

He returned to the minivan and opened her door to help her out, but MaryJane lurched the other way, toward the driver's side. When he reached for her, she slapped his arm out of the way, pushed past him out the door, and walked to the elevator. She was still barefoot. Longo grabbed her hiking boots and caught up with her.

They rode the elevator to the second floor and walked the hallway to Unit 211. As Longo was preparing to unlock the door, he noticed that it was ajar—the door was difficult to properly close, and MaryJane had left it that way before. He'd asked her to be mindful of this, and was exasperated that she'd done it again. He pushed the door open and entered the condo. MaryJane stood in the hallway, refusing to come inside. She'd begun to weep, Longo said, and was mumbling incoherently. "That's when I started to get alarmed," he testified. "I knew something was wrong."

She started moving back down the hallway, toward the elevator, and Longo grasped her around the waist, carried her into the condo, and shut the door. Her hysteria immediately heightened. Longo let go of her, and she slumped to the floor. "I'd never seen emotion like this," Longo said. MaryJane, he pointed out, had always been a quiet and rational person.

He was concerned that something had happened inside the condo, so he made a quick tour, running from room to room. Nothing appeared out of place. He didn't see the kids, but he figured that MaryJane had simply brought them over to Denise Thompson's house. "The only thing I noticed," he testified, "was the kids' stuffed animals were sitting on the couch, which alarmed me somewhat, because if they went over to Denise's they would take their stuffed animals with them. These were things that they would carry with them all the time."

By now, Longo had lost his temper. He began to yell. "What's wrong?" he shouted, but there was no articulate response from MaryJane. "She was literally on the floor, curled up in a ball, bouncing back and forth, hitting her back against the wall, crying, wailing, moaning—sounds I've never heard come out of MaryJane, or actually anybody else," he testified.

Longo continued yelling, but she wouldn't answer, so he raced through the house once more, this time in a state of panic, switching on all the lights. He looked in the bedroom, and there, on the unmade bed, nearly hidden between the pillows, he saw his youngest child.

At first, he felt a wave of relief—at least one of my kids is okay, he thought. But when he bent over Madison, he noticed that her skin looked purple. She wasn't moving. He jostled her. "She was extremely cold to the touch," Longo said. "I feared for the worst." MaryJane was still at the condo's entryway, still in hysterics. He thought she might have done something to Madison, and his anger surged. "That was the first time I had any sense of wanting to use physical force," he said.

By this point in his testimony, everyone in the courtroom was motionless. Even Briggs, who'd seemed perpetually caffeinated, had ceased rocking in his chair. The jurors sat with their backs angled forward; Judge Huckleberry kept a hand curled over his mouth. Longo delivered his testimony in an unwavering voice. Only his posture, which for the duration of his time on the stand had assumed a soldier's alignment, finally began to deflate.

In the condominium, his panic escalating, Longo left Madison on the bed and returned to MaryJane. His wife, he testified, remained unresponsive, tucked into herself, lying on the floor. "I'm asking her, 'What's wrong? What's wrong with Madison? What happened?'" Again there was no reply, and when Longo knelt beside her, she lunged at him with her fists, pounding on him, swinging wildly. "I finally ended up grabbing onto her robe and just lifting her up against the wall and just said, 'You have to control yourself.'"

She struggled to break free, but Longo dragged her into the bedroom. When he released her, she again collapsed to the floor, shrieking. Madison was on the bed. There was no sign of Zachery or Sadie. He asked her where the other children were, but she did not answer. Longo lifted her off the floor and pinned her against the wall. "I shook her against the wall pretty violently trying to snap her out of it," Longo testified. "I ended up hitting her head probably a few times against the wall until she finally came to some sort of sense. She calmed down a little bit."

He tried to speak with her. "You've got to tell me what's going on," he said, moderating his tone, trying to pacify her. "Where are Zachery and Sadie? What's wrong? You need to tell me."

At this moment, Longo testified, she seemed fairly lucid. She even looked at him. Longo let her go, and she remained standing. She started to speak. "You did this," she told him. "This was your fault." Longo replied that he didn't understand what she was talking about. "You did this," she said again. "You killed us."

Her use of the word "killed," Longo said, caused him to lose all restraint. He was terrified that all three of his children were dead. He grabbed MaryJane again and banged her head against the wall. She kept repeating, "You did this. You killed us."

"She wouldn't stop saying that," Longo testified. "And I couldn't stop her from saying that." He pushed her against the wall once more, and she fell to the floor. "After that she was just on the ground. She wasn't mumbling, she wasn't crying. She wasn't really doing much of anything. She was trying to cover her eyes. I was yelling as loud as I've ever yelled before at her. I was yelling, 'Where are they? I want to know where they are. I need to find them.' That's when she said something about they're by the house. 'They're in the water. They're by the house.' That's when I lost it."

With one hand, he grabbed the lapels of MaryJane's robe and hauled her to her feet. The other hand he placed around her neck. "And then I just started squeezing," he said. She grabbed him by the forearm, he said, but he kept choking her. "I lost her at one point, started to drop her. I grabbed her with both hands and continued to squeeze. And I didn't stop for a long time. I didn't stop until I couldn't hold her up anymore. I let her drop to the ground."

He ran out of the bedroom and into the living room. There, he says, he collapsed onto the floor. He lay on the carpet for some time, he's not sure how long. The house was silent. He thought of calling the police, he said, and even picked up the telephone, but never dialed. Instead, he had a momentary delusion—he believed that all he'd done could be corrected. MaryJane, he felt, couldn't have been too badly injured. "I thought that I could fix it, that everything would be okay," he testified. He returned to the bedroom. MaryJane was still there, in the same spot. "That's when I knew that she was not getting up," he said.

Longo's voice, for the first time during his testimony, cracked with emotion. It rose an octave; it was punctuated with rapid breaths. He wiped his forehead, twice, with the back of his hand. "That's when I knew," he said, "that she was dead."

His first idea, he testified, was to hide her. He didn't know where. He looked around the bedroom. The closet door was open; inside it were several suitcases. He brought one out and opened it on the bed. He lifted MaryJane off the floor and carried her to the suitcase. Her bathrobe slipped off as he moved her. He bent her limbs so they'd fit inside, and he zipped it closed.

He opened a second suitcase and prepared to do the same with Madison. But the suitcase was large and his daughter was tiny. "I didn't want to put Madison in a suitcase like that," he said. "I ended up trying to make it more comfortable." He pulled out the bottom drawer of the bedroom dresser, the one with all of Madison's clothes, and dumped the contents into the suitcase. Then he went over to pick up Madison, and he saw something that startled him.

Her chest rose up and sunk down. He was stunned. He'd been certain she was dead. But as he watched her, he saw her chest rise and fall, rise and fall. He shook her, but she did not stir. "I didn't know what to do," Longo said. "Even though she was breathing I thought of her as dead at that point. There was nothing I thought that I could do to make her responsive. I couldn't put her in a suitcase like that, though, as she was still alive to some extent, but she wasn't alive."

Longo spoke these lines quietly but clearly. He appeared calm. He rubbed the right side of his face a few times, as if expecting to wipe away tears, but his cheeks remained dry. I later asked him, over the phone, how he was able to keep his emotions in check at such a moment. "I've been through it so many times in my mind," he told me. "I just try to separate myself from it."

Throughout the courtroom—in the spectator section, at the lawyers' tables, in the jury box, and with Judge Huckleberry himself—jaws were clenched; eyes were either pinched shut or held wide open; palms were pressed against cheeks. Even Steve Krasik wasn't speaking anymore. When Longo had launched into the description of the final moments of his family's life, Krasik had asked the judge if Longo could simply continue his account, without the need for questions, and Huckleberry had allowed it. So Longo kept talking, uninterrupted, to a stunned and silent room.

"I ended up putting my hand on her throat," he said. "To cut off her air supply. She seemed to not breathe instantly. I let go. I saw her breathe again. I put my hand on her throat and squeezed, until I knew she couldn't breathe anymore. I put her in the suitcase."

He carried both suitcases out of the condominium, into the elevator, and across the parking garage to a set of stairs. The stairs led up to the boardwalk. It was still before dawn on Tuesday, December 18, 2001. Nobody was around. He turned left on the boardwalk and then right down a gangway that led to the docks. He continued past a set of fish-cleaning tables and reached the edge of the water. He dropped the two suitcases into the bay. He didn't learn the details of what happened to Zachery and Sadie, he said, until after he was arrested in Mexico three and a half weeks later.

THIRTY-SEVEN

AS I SAT in the courtroom, with a pen in my hand and my notebook open on my lap, listening with almost trancelike intensity to the culmination of Longo's story, I was able to conjure only a single word. I scribbled it in my pad, in giant letters, in the center of the page. I circled it. I added an exclamation point. I wrote: BULLSHIT.

He'd lied. I was sure of it. But it wasn't so much the lie that repulsed me. I'd known, all along, that Longo was an able and willing liar. What shocked me was the nature of the lie. It was an ugly lie; an evil lie. Longo had just announced, out loud, in public, that MaryJane was the real killer—an irrational, uncontrollable, cold-blooded killer. She was initiator of the crimes. She had either murdered or tried to murder all of her children. Longo had said this even though, in all his testimony, he'd never provided a moment of insight into what might cause his wife to act this way.

Longo had told his story in front of two of MaryJane's sisters and her sister-in-law. I saw only Penny's face, etched with rage, before the three of them darted from the courtroom. I was mortified that I'd affiliated myself with Longo—that I had actually cared about him, had wished for him the most humane possible punishment, away from death row.

What could possibly be the point, I wondered, of such a lie? Maybe he wanted to show that he was nothing more than a gentle, loving man who had been driven to murder by overwhelming circumstances—that he was as much a victim of these crimes as he was the perpetrator. Perhaps he hoped to foster reasonable doubt in at least one juror, and thereby save his own life. But no one, I thought, could possibly accept his testimony as truth.

Steven Briggs, of course, didn't accept it. After a short break, he launched into his cross-examination of Longo. It was a riveting performance. Briggs sat on the edge of his chair, palms flat on the armrests, elbows out, the muscles in his forearms visibly twitching, as though he were on the verge of leaping from his seat. His tone was sneering, sarcastic, belittling.

Longo managed to maintain his self-confidence, though, it seemed to me, just barely. In any case, almost everyone in the courtroom was watching Briggs—the look on his face was of a boxer one blow from a knockout. More than once, until he was reprimanded by Huckleberry, Briggs asked a new question before Longo had finished with the previous one, as if Longo's answers weren't worth listening to. What was important was the sheer breadth of questions raised by his testimony.

"Mr. Longo," Briggs began, "you've been on the witness stand now over the course of about four days. Is there anything else you'd like to tell us?"

"Not that I can recall," Longo said. "But there's a lot of stuff that happened in the course of a couple of years that I would want to air if we think of it."

"MaryJane's not here to tell us about your past conduct, is she?"

"No, she's not."

"She's not here to tell us how you really treated her and the kids?"

"No, she's not."

"She's not here to tell us what happened that night, is she?"

"No, she's not."

"She's not here because you killed her."

"That's true."

"All we have is your word, right?"

"That's correct."

"At some point," Briggs continued, starting to deconstruct Longo's story, "you decide that you're going to tell MaryJane about all your past crimes?"

"Yes," Longo answered.

"You continue to reveal thing after thing after thing?"

"Extremely reluctantly. It took several hours for this to take place."

"At some point you revealed too much. She got upset with you?"

"Yes."

"Essentially kicks you out of the bedroom?"

"Yes."

"You go to work later in the day?"

"Yes."

"MaryJane, the next thing you know, she has come back to pick you up?"

"Yes."

"In her state, her mental state," Briggs said, "she still remembers to come to Fred Meyer at eleven o'clock to pick you up?"

"Yes," said Longo. "I actually wondered if she would. I was concerned about that."

"Then you arrive back home at your condominium, and you see Madison. You think Madison might be dead?"

"Yes."

"And you don't do anything for Madison. You shake her?"

"Yes."

"You don't call for help?"

"I was hysterical at that point."

"You were hysterical?"

"Yes."

"You don't call nine-one-one?"

"No."

"You don't call the police?"

"No."

"Don't call the fire department?"

"No."

"MaryJane says to you, 'You did this. This was your fault. You killed us.' Things along that line?"

"Yes."

"Your response is to strangle her?"

"Yes," said Longo. "This was a good forty-five minutes after we'd been home."

"Put her naked body into the suitcase and zipped it closed?"

"Yes."

"Then you went to Madison?"

"Yes."

"And you go to pick her up and you see her chest move. She's still alive?"

"Yes."

"And you don't call nine-one-one? And you don't call the fire department? And you don't do anything?"

"No."

"You strangle her. And you strangled MaryJane based upon her hysterical statements and the appearance of one child who turned out to be alive, according to you. And then you pack up all the clothing to dispose of, everything that was in the condo?"

"Eventually, yes."

"So you spend time going to at least one dumpster location, maybe two dumpster locations. But you don't go out and look for your other two children?"

"Actually, I did," said Longo. "I drove all around the Newport area, all around the bay. I went halfway to Waldport and decided that if they were in the water somewhere, as MaryJane had said, then it's too late. There's nothing I can do."

"You don't ask anybody to help you in your search?"

"I considered, actually, calling Denise to find out if the kids were over there."

"You didn't call Denise?"

"No, I did not."

"You didn't call anybody?"

"No, I didn't."

"You waited," said Briggs, "until Starbucks opened and you went and got some coffee?"

"Yes," said Longo.

"You didn't know what had happened to Zachery and Sadie?"

"I did not know for a fact what had happened, no."

"You didn't know that there were any rocks tied around someone's ankle?"

"No."

"You didn't know whether they were alive when they went in or dead?"

"I still don't know."

"You dispose of MaryJane's body and Madison's body?"

"Yes."

"Steal a car?"

"Yes."

"Go to work?"

"Yes."

"You begin to tell everybody that your wife is an adulterer?"

"Actually, I thought I had said it before this point."

"That she left you for a guy who made more money?"

"Yes."

"You get off work and you go to the video store?"

"Yes."

"And you rent a movie called *Blow*?"

"Yes."

"That's a movie about a guy who deals drugs. Johnny Depp, I think, is the star of it, right?"

"Actually, you know more than I do. I never ended up watching it."

"All you saw was how it was marketed?"

"Yes."

"So that's the movie you decide to rent to keep your mind off the murder of your family. Is that right?"

"Yes, actually it is. I can't argue with that."

"You go to the gym to work out?"

"Yes."

"You go in and get your paycheck on that Friday. Drive to San Francisco, correct?"

"Correct."

"Ultimately you fly to Cancún?"

"Yes."

"Adopted the identity of a *New York Times* writer?"

"That's correct."

"Went dancing?"

"Once, yes."

"Bought dinner for Janina?"

"Yes."

"Shared a cabana with her?"

"Yes."

"Planned to travel south to Guatemala?"

"Yes. I had one week and a half left of my stay. That's what the plans were."

"You're telling us that your plan was to come back within the next week and a half so you could fly back to San Francisco?"

"That's correct," said Longo.

THIRTY-EIGHT

LONGO FINISHED TESTIFYING late in the afternoon on April Fools' Day, 2003. The next morning, the defense rested, and on the following day, both sides presented closing arguments. Paulette Sanders recounted Longo's history of lies and deceit and questioned why, with his life on the line, he would now suddenly start telling the truth. "What did he have to lose?" she said. "He's going to tell one more story."

Steve Krasik pointed out that the bodies were found in two different places, twelve miles apart, and that this indicated there were two different killers. If Longo really wanted to lie, Krasik added, he would not have admitted to any murders. He'd have proclaimed his complete innocence. "This is not the story we would be telling if we were working on a story," Krasik said.

The court then recessed for a three-day weekend. On Monday, April 7, Judge Huckleberry issued a brief set of instructions to the jury and sent them off to deliberate. The twelve jurors filed into a private room; Longo was escorted back to jail; and I waited, along with a few other members of the press, in the hallway outside the courtroom. I didn't want to risk missing the verdict. So I sat there and fed myself junk food from the vending machines and read and reread Longo's newest letter, the first he'd sent me since his trial had begun.

He had mailed it to my rental house in Newport the day after he'd completed his testimony. It was a strange and tortured letter, by far the most convoluted he'd ever sent me. I had to read it three times, while taking notes on a separate sheet of paper, in order to grasp the letter's internal logic. Over the course of thirty-seven pages, Longo attempted to explain all of his actions from the day of the murders, in December of 2001, to the final moments of his testimony, sixteen months later.

First, though, he expressed his hopes that we could salvage our relationship. He was acutely aware that I was retreating from him. When Longo was leaving court after being cross-examined by Briggs, we'd briefly made eye contact, but I had felt so disgusted with him that my instinctive response was to quickly turn away. Longo referred to this reaction on his letter's opening page. "I do hope that I haven't lost you as a friend," he added.

Then he began his explanation. What he'd said on the stand, he insisted, was entirely true: MaryJane really did kill Zachery and Sadie, and attempted to kill Madison. Her objective, Longo wrote, was to hurt him as badly as she could. That's why she killed the kids—to punish Chris. "I'm sure most people don't see that," he said.

He told me that he was infuriated with MaryJane, and that was why he killed her and then slandered her by concocting the story of her adultery. After Zachery's body was found, he fled Newport because he, like MaryJane, was a murderer, and he did not want to spend time in jail before he had a chance to properly grieve.

During the drive to San Francisco, he wrote, he had an epiphany. "I rapidly came to the realization that I truly was to blame. I sent her over the edge. Without me & my lies, none of this would have happened. This was all my fault." These insights, he added, allowed him to forgive MaryJane.

He traveled to Mexico, he wrote, because it was inexpensive there, and his remaining money could last several weeks. "I had intended for it to be a period of clarity & a time to deal w/ the grief of the situation, before returning to deal w/ the legal end." Once he arrived in Mexico, though, he "wanted to think of nothing," he wrote. "I had fun."

After he was arrested and jailed, he at first insisted he was innocent of all the crimes. He told his lawyers and me that he "wasn't at home" when his family was killed. His lawyers, of course, needed more than this to work with, so he provided them with a detailed story.

What he said was that, on the night he came home late from work and drank his wine and ate his cheese, he did not lie down next to his wife. He went for a drive. He parked the minivan at an oceanside overlook and dozed for several hours. When he arrived back at the condominium, an intruder was in the living room. "The guy," wrote Longo, "looked crazed & out of it." No one else seemed to be in the condo, and Longo became alarmed.

"Where the hell is my family?" he yelled.

"The bitch wouldn't listen and now she's gone," the intruder said. "They're crab bait, in the bottom of the bay."

The man then attacked him, and Longo grabbed a clothes iron and struck him in the head with it, killing him. Longo dumped the intruder's body in the water and, knowing that he couldn't go to the police without facing arrest himself—he'd just killed a man—he stole a car and soon flew to Mexico.

Longo described this as "a plausible story." Later, I asked Krasik what he thought. "It would have been more believable," Krasik told me, "if he'd said a UFO came down and abducted his family."

Over the course of the year Longo spent in the Lincoln County Jail, he resolved to transform his life. He decided that, when it came to his trial, he would be "completely & utterly honest in every aspect." So he confessed to the two murders he committed, then took the witness stand and told the truth about the rest.

Within the framework of Longo's letter, "the truth" meant that

MaryJane was a murderer. But after he told this truth on the stand and saw my reaction, and then heard from his lawyers that his story did not seem to go over very well, he realized that he may have made a mistake. He had long ago forgiven MaryJane for her role in the crimes, and he understood, too late, that it may have been more honorable for him to accept responsibility for all the murders and ensure that his wife was remembered as a kind and gentle woman.

"Maybe this is the one time that I should have lied," he wrote. "Maybe in some way I took the cowardly course by not taking all of the blame. Maybe for the second time in the last year & a half, I've made a horrendous & devastating decision in telling the absolute truth—the first ending in the deaths of my entire family."

If he hadn't told the truth to MaryJane that night, he reasoned, then she and the children would probably still be alive. "My road to honesty," he wrote, "may be proving more treacherous & debilitating than the egregious course of lies & dishonesty."

At the end of his letter, he repeated once again that he did tell the truth on the witness stand—"I did not take the lives of Zack & Sadie"—but he added that he now felt ashamed of himself for being honest in court. His final conclusion was this: "I'm confused even more now, to the point of whiplash."

Well, the jury was not confused. Four hours after Judge Huckleberry had dismissed them to deliberate, they returned a note stating that a decision had been reached. I was still sitting in the hallway. The first indication that something had happened was the command, crackling over a court officer's radio, to "clear the halls." The halls were promptly emptied of people, and a minute later Longo emerged, blank-faced, swinging his left arm, leading a small procession of officers. My heart rate quickened and my stomach knotted, and I ducked away before he could see me.

I entered the courtroom behind most of the other spectators and ended up taking a seat in the back. Up front, where MaryJane's sisters and Joe and Joy Longo sat—the two families, so far as I could tell, were not on speaking terms—a box of tissues was passed along the row, and everyone took a few. Huckleberry entered and said, "Well, it looks like we have a verdict," and he remained standing as the jury filed in. Then the judge turned to face the spectator section and asked us to please maintain decorum and refrain from any outbursts while the verdict was read.

Huckleberry asked the defendant to stand, and Longo rose, along with Krasik and Hadley. Longo kept his hands clasped behind his back as the judge read from the jury's verdict form. Huckleberry began with the findings in Zachery's murder. "We, the jury," he read, "being duly sworn and impaneled, find the above named defendant guilty of the charge of aggravated murder."

I stared at Longo's back but detected only the slightest movement, just a quick curl of his fingers. Penny and Sally clutched at each other, arms interlaced, and had already begun to weep by the time Huckleberry revealed that the jury had also found the defendant guilty of murdering Sadie. The jury foreman later said that reaching a unanimous decision had not been difficult. Though the jurors had remained in the deliberation room for four hours, nearly everyone, the foreman said, had agreed on Longo's guilt within minutes.

Longo called me that evening. He said that he was about to be locked down for the night, and had only a short time, but he felt the need to talk. He was relieved, he told me, that I'd picked up the phone—he thought the guilty verdicts may have permanently ended our conversations. "I didn't know if there was going to be any more," he said.

I asked how he was feeling. "I'm definitely not in a good mood," he replied. While he was waiting for the verdict, he said, he'd remained hopeful. But when the officers came to bring him back to court, he knew the decision was likely to be against him. It had been reached too quickly. The one person who'd tried to encourage him, he said, was the officer walking directly behind him, escorting him into court. "He was just like, 'Well, you never know what's going to happen. You know juries go one way or another and you can never tell.'"

I didn't have much to say. I had tried to decipher Longo's system of reasoning, but had ended up completely disoriented. I told him that it might be better if we didn't speak on the phone for a while. "Let's communicate by letters," I suggested. I explained that witnessing his trial had been an unsettling experience for me—"I've just been having a real tough time with it, Chris," I said.

"I understand," he said. He added that the trial was almost over; all that was left was the penalty phase. "It should be quick and painful," he told me.

The penalty phase lasted four days, during which Joe, Joy, and Dustin all testified. "I hate what he did," his father said, "but I still love him." Dustin expressed similar sentiments. He added that he'd spent much of his life in awe of his older brother.

When Joy sat in the witness box, she appeared to focus on some empty place in the center of the courtroom. She shook her head slowly back and forth as she mentioned the note she'd written to Chris, pleading with him not to get married. Then, when she spoke of Longo's disfellowshipment—"It was a form of discipline," she said, "that was necessary for Chris"—she finally glanced over to her son, and Longo gazed back at her, and for a few seconds everything in the courtroom seemed to halt. Neither Chris nor Joy's face really changed. They just looked at each other, motionless and stoic, as if the helplessness of the situation had overwhelmed any reaction.

Ken Hadley, in a speech to the jury, appealed for compassion. "We're going to ask you to believe that there is justice without killing someone," he said. "We've had enough killing." He pointed out that before the murders, Longo had never displayed any violent behavior. "Whatever caused this horrible thing to happen," he said, "is not a pattern in his life."

Briggs delivered a brief rebuttal. "Four innocent people are dead and one guilty man is alive," he said, chopping at the air with his hands. "There's an injustice as we sit here in this courtroom, but you have the power to correct it." Zachery, Sadie, and Madison, he added, were robbed of their entire lives. A parent's greatest fear, Briggs concluded, was having a child die, yet this was something Longo actively sought: "He wanted his children dead."

Judge Huckleberry briefly addressed the jury. "There's a lot on the line," he cautioned. "Any one of you has the power to choose life in prison as the sentence." Then he sent them off to deliberate once more. The next day—Wednesday, April 16, 2003—they informed Huckleberry that they'd reached a decision.

Everyone reassembled in the courtroom. Upon the judge's command, Longo stood. If he was nervous, he didn't show it, as usual, he managed to appear wholly unconcerned. His fingertips rested lightly on the defense table. Hadley's arms were crossed on his chest. Krasik's hung by his sides. Joy closed her eyes, though this did not prevent a couple of tears from leaking out. Joe looked down at his lap, as if in prayer. The jurors were dry-eyed and passive; a couple of them were chewing gum.

The judge read the sentence. He began with the first count, the killing of MaryJane. "Should the defendant receive a death sentence?" Huckleberry asked. He paused and glanced at Longo. "To this question, the jury has answered, 'Yes.'"

Longo bowed his head and grimaced slightly. Joy exhaled, but didn't cry anymore. It was as if all the energy had finally gone out of her. Hadley covered his face as Huckleberry continued reading. For the murder of Zachery—death. For the murder of Sadie—death. For the murder of Madison—death.

Afterward, Huckleberry addressed Longo. "The facts of this case reach a level of perfidy beyond anything I've experienced in my life," he said, speaking in the sort of scarcely restrained whisper that feels more intimidating than a shout. It was a tone he hadn't previously used during the trial. "The sheer breadth of harm truly makes it impossible, in my judgment, for you as a person to either atone for these crimes or expect absolution," he continued. "I do not know how the scales could ever, ever be leveled."

He placed Longo in the custody of the Oregon Department of Corrections, for transport to death row at the state penitentiary. An officer then cuffed him, behind his back. It was the first time he'd been manacled in court, and Longo, now a condemned man, was marched away.

THIRTY-NINE

I REMAINED IN the courtroom as Joe and Joy filed out, and then Penny and Sally. It was clear, glancing at their faces—shocked, uncomprehending—that this wasn't the sort of trial whose verdict would ease anyone's emotions. No one was celebrating. The crime had been too enormous. Police photos of their loved ones had been publicly displayed; details from the autopsy reports had been read aloud. I felt terrible for them. "Many people talk of closure," wrote Penny Dupuie in a statement that she later distributed to the press. "There is no closure. MaryJane, Zachery, Sadie, and Madison are gone."

After the spectators and lawyers and officers had all left, I still didn't move. Soon I was alone, except for a television cameraman who was gathering a length of cable. I felt pinned to my seat, exhausted and head-heavy. It seemed as if my relationship with Longo—from the *Oregonian* call, to the Wednesday talks, to the flood of letters, to the final verdict—had happened so rapidly, one event atop the other, that it was as though a coil had been compressed in my mind, and now, with the end of the trial, everything had sprung loose. And as I sat there, on the wooden bench inside the courtroom, no longer anxious to learn the whole of Longo's saga, no longer consumed by the quest for a story, I realized at last exactly how I felt about him.

I hated him. I hated him in the intense way that you can only hate someone you'd once truly cared about. What Longo had done to his family, and how he had acted in court, and the ways in which he'd toyed with me, and the fact that he had never expressed any genuine remorse or even seemed to grasp the magnitude of his crimes—I saw all that and I hated him more. I was thankful for the jurors' decision. I agreed with them. Longo, I thought, deserved to die.

He had come to me as a liar and con man and possible killer, and during our time together I had given him every benefit of the doubt. I had accepted him as at least a partially kind and empathetic man; I had believed that he was striving to become more honest and trustworthy. But when it came to the ultimate test of his essential decency—in court, under oath, in front of his and MaryJane's families—he told the biggest and worst lies of his life.

I had been fascinated by Longo. I'd also been fooled by him. As he was led out of court for the final time, he seemed to me not much different than the day he'd first called. He left as a liar and con man and definite killer. He was gone, condemned to die, and I had this sense of having survived something—a storm of sorts, and here I was on the back end, alive and intact, though in many ways not the same person at all.

It was a twelve-hundred-mile drive back to Montana. I spoke to Jill on my cell phone for hours, but even she couldn't unjumble my head. I was furious with Longo, and I couldn't stop thinking about him. Jill said that this was understandable. The time I'd spent with him, she said, had been too intense to simply come to a clean and sudden close.

She was right; there was unfinished business between Longo and me. Toward the end of the trial I had backed off from him, but I had never let him know exactly how I felt about his performance on the witness stand and his continued insistence that he was not the only killer. I thought that telling him this, whether he cared to hear it or not, would settle my own mind.

When I returned home, I wrote him a letter. It was ten pages long, badly rambling in parts, but I tried to express some of my feelings:

> Chris, when I sat there in the courtroom on the day you told the story . . . I felt this strong and horrible feeling in my chest. . . . I felt sick, physically. . . . Your story did not feel real to me. It felt awfully wrong. . . . I know that you don't want MaryJane thought of as a crazy and evil murderer, but I also know that you just weren't ready, at your trial, to speak the deep, dark truth. You gave a whole speech at your trial about coming clean, but you did not. And that makes me more furious than I've ever been at anyone my whole life. . . . To blame it on your dead wife is quite evil. . . . If something feels false, tastes false, smells false, sounds false, and if there was a guy on a bridge at 4:30 in the morning who saw a red minivan, and if there are all the reasons in the world for you to tell a false story (so people don't think you were capable of such a heinous crime without proper provocation) and if there is not a single bit of evidence to back up your story . . . then Chris, come on, what are we all supposed to think?

I mailed the letter to the Oregon State Penitentiary. It seemed like a fair epitaph to our relationship—or at least it felt good to write it. I didn't care if Longo responded. In fact, I hoped he wouldn't. I hoped it was the end.

FORTY

IT WASN'T. A week later, I received a thin white envelope in the mail, different from the brown envelopes Longo had used in county jail. The return address identified him as inmate number 145-09-855 of the Oregon State Penitentiary. Inside was an eight-page letter, written in blue ballpoint pen. Apparently, he was no longer forced to use golf pencils.

"I feel horrible that you are going through all of this w/in yourself," he wrote. "I'm truly sorry." This response was predictable; Longo was good at expressing generalized contrition. What surprised me was the next part of his letter.

A month earlier, during one of the last phone conversations I'd had with Longo—this was toward the end of his trial, just after he had finished with his testimony—we'd had a rather blunt exchange. "I have a feeling," I'd said, "that if you're strapped to a gurney and they're about to inject you—sorry to be graphic—and I say to you, 'Well, you're about to die, do you want to change anything that you said in court?' I have a feeling you will say no. Is that a true feeling?"

"That is absolutely true," Longo had said, and I was convinced that he would never budge from this position.

But now, in the first letter he'd sent me from death row, Longo completely altered his story. "What I said on the stand was false," he wrote. "I am absolutely guilty of killing my entire family."

Longo then proceeded to explain precisely why he'd lied in court. The whole trial, he wrote, was a sort of suicide-by-jury. He was *trying* to be put to death. He had it all mapped out. "I'd admit the past & monsterize myself in the eyes of the jury," he wrote. "I would try to be emotionless, to add credibility to that monsterization. I would tell the story as planned to cement the hatred of both loved ones & the jurors, which would guarantee the guilty conviction & pave the way for a death penalty decision."

This was, he noted, a performance of utmost gallantry: Ensuring his own death was the best thing he could do to assuage the grief he'd brought to his and MaryJane's families. Everything had worked out perfectly. "Mission accomplished," he wrote.

He even chided me for not figuring this out myself. There'd been no need, he added, for me to get so upset—clearly, I'd taken the trial way too seriously. He'd really meant it all as a grand charade. "I know that you didn't believe me on the stand," he wrote. "I know that nobody else believed me either. What I said in regards to what happened that night wasn't meant to be believed. It was meant for everyone to despise me & move on." He poked fun at Krasik and Hadley as well. "My attorneys," he snickered, "believed it to some degree."

Besides, he added, he didn't want to live in the general inmate population, anyway. "Being out of my cell for twelve hours a day & in the 'world,' the population, didn't exactly fit my needs," he noted. He desired a cell on death row. "I still have a ton of introspection to do & I don't know that I could accomplish this from anywhere but here."

He concluded the letter with an expression of gratitude. "You've had, & are continuing to have, a tremendous impact on my own life," he wrote. "No one has spoken to me w/ quite your level of honesty. You're a great example for me. Go figure. Two liars to make two people turn to a path of honesty."

* * *

I wasn't sure how to react to this letter. It seemed to belittle our entire relationship, to annul the twenty-four letters that came before. Longo had spent more than a year trying to convince me that he wasn't a monster, only to inform me that what he most wanted was to be seen as one.

His words were confounding. Did he really believe that the trial operated under his command? That he had power over all twelve jurors, both his lawyers, and Judge Huckleberry? The reasoning he displayed in this letter—"I made the decisions that I did quite consciously"—led me to think that Longo may have descended into madness.

Then I thought that this was exactly what he wanted me to think. His trial had ended with the worst possible outcome, and now he was counting on me, the person who supposedly knew him so well, to write that he'd lost his mind. Perhaps he hoped that, after my account was published, he could use the appellate process to gain a new trial—one in which he'd plead not guilty by reason of insanity. Or was this idea itself insane?

I had no clue. The only thing I knew was that I didn't know enough. "If it's important to you that you know everything that truly happened, we can talk about it," Longo wrote toward the end of his letter. "There's still a lot that we have to say; at least on my end." This time, he implied, he'd tell me the real, true story.

I couldn't resist. I sent him another letter. "Yes, Chris," I wrote, "I'd like to take you up on your offer: Please tell me the full story of what happened."

And so, once again, he described the murders. Everything he said on the witness stand was true, he claimed, until the very end. He did confess all of his lies to MaryJane; he did have an argument with her; he did go to work the following afternoon. On the final day of his family's life, though, Longo drove himself to the Fred Meyer. Mary-Jane did not take the minivan, and therefore she did not pick him up.

A little after 11 P.M. on Monday, December 17, 2001, Longo drove himself back from the department store. "I came home to a silent condo," he wrote, "went directly to the bedroom to check on MJ, & found her holding a pillow over Madison, pressing it over the upper half of Maddy's body, crying, kneeling on the bed."

As I read this, I felt a familiar sense of dismay. He was still insisting that MaryJane had initiated the murders. I told him, in my next letter, that it was impossible to believe that he happened to walk into the condominium at the very moment his wife was in the midst of a murderous act. "You were gone all day, Chris—if MJ wanted to harm any of the kids she had hours and hours to do it," I wrote. "Is it possible that, in the trauma following the crimes, your mind may have substituted one image for another? Is it possible that, when you came home from work, MJ was leaning over Madison, changing her diaper?"

"I personally don't believe that it was sheer coincidence that MJ was in the middle of the act at the exact moment that I arrived at home," Longo responded. "I really believe that she was doing this to spark a reaction from me. What kind of reaction she expected, I don't know." He was also confident that his mind wasn't playing tricks on him; MaryJane's actions, he insisted, were unmistakably aggressive.

"I realize that it just doesn't make sense that MJ would ever do such a thing," he continued. "I know that no one, including to a large degree myself, would ever think her capable. But I would never think that I was capable either; that I would ever conceive of doing such a thing, right up to that evening when I walked into the condo."

Once inside, Longo killed MaryJane in the manner he had described on the witness stand, except that there was not as much banging and yelling. Then he strangled the breathing but unresponsive Madison. He described the feeling that came over him as he killed his wife and youngest child as a kind of feral surge, one that short-circuited his brain's ability to reason.

Throughout the first two murders, Longo wrote, Zachery and Sadie remained sleeping on the sofa bed. They awoke only after Longo had thrown the suitcases containing MaryJane and Madison into the bay. He scooped up his remaining two children and carried them to the minivan. "I had no plan or idea of what I was going to do, only that I wanted us to be somewhere else," he wrote. He drove off in the van, and Zachery and Sadie soon fell back to sleep.

He headed south, following the coastline. Longo thought about what he would tell his children when they awoke. This notion filled him with terror. "I resolved that Zack & Sadie would suffer less if they weren't alive," he wrote. He stopped the minivan in a residential neighborhood and picked up two large rocks from in front of a house. Then he drove some more—"looking for a spot," he wrote. He ended up in Waldport, on the Lint Slough Bridge. It was about 4:30 A.M. on December 18, 2001. Longo conceded that he did indeed meet Dick Hoch, at the very time Hoch testified to, only one day later.

He couldn't recall if he met Hoch before or after the final murders, but he did remember some of his actions. "I put a rock in Zack's pillowcase that was on the floor of the van, tied it around his ankle while he was still asleep. He barely stirred. I picked him up, blanket & pillowcase, & threw him in on the south side of the bridge, where I pray that he never fully woke up or knew what I was doing to him. At the time I really felt like I was doing the honorable thing, the best thing that I could do."

As for Sadie, his recollections were vague but horrifying. "I have dreams of Sadie screaming," he revealed. She's falling away, dropping into the water, crying out. "I can't tell you," he wrote, "how much I hope & try to convince myself that that's not what happened."

Longo had finally admitted to me, in June of 2003, eighteen months after the crimes and fourteen months after our first conversation, that he'd murdered his entire family. He still claimed that MaryJane had instigated the killings, but by now I realized that all I had to do was write him a few pointed letters and he might very well recant this part of his story as well.

I didn't write those letters. I didn't because of an idea I was unable to push out of my mind: Just because a liar says something you want to hear doesn't mean it's true. I could not ignore the possibility that his confession to all four murders was also a lie. Longo, I knew, would lie about almost anything if he thought he had something to gain.

What could he gain by admitting to two more murders? He may have thought that by doing so he could reestablish our friendship. In the letter Longo had written after his testimony—the one I'd read while waiting for the verdict in the guilt phase—he had described his feelings about my decreasing presence in his life. "You've given me a support that I never expected & it has meant more to me than the tacit support of my parents & brother," he wrote. "But now I'm afraid that you are beginning to draw that line. I don't mean to sound as though I don't deserve to be an outcast, I just don't want to lose so much."

Keeping me as a friend, he continued, would allow him to "save a little piece" of his life and give him "something to hold on to." "The possibility of total loss is overwhelming," he added. "It's making me wreckless." When he wrote this, Longo was still insisting that MaryJane had killed Zachery and Sadie. But he hinted that he'd be willing to change this story. He was thinking, he said, of issuing a fuller confession, even if he had to admit to things he didn't do. "I wonder," he added, "if that news would bring you back to a greater level of trust in me."

Now, a few weeks after he'd been incarcerated on Oregon's death row, he had made such a confession. But I didn't move to a greater level of trust. I only wondered why Longo had suddenly admitted to all four murders. Were we inching toward honesty, or was he simply saying whatever he thought necessary at this point to placate me? The question was more or less rhetorical, for even if I'd posed it to Longo, his answer wouldn't matter—any response might be another lie. It would have been just as good to flip a coin: heads, he's telling the truth; tails, he's placating.

I realized, then, that I couldn't be certain of almost any aspect of the murders. I was planning to write a book about the crimes, but the truth is, despite all I'd seen and heard and read, I wouldn't feel comfortable definitively answering any of the five basic questions that should be addressed in the opening paragraphs of a half-decent newspaper article.

Who killed them? Longo has never changed his contention that MaryJane was involved in at least one of the murders, and there are no living witnesses or forensic evidence to prove him wrong.

What was the manner of death? The prosecution maintained that all four victims were strangled; they said the autopsy reports supported this presumption. Longo insisted that MaryJane and Madison were strangled and Zachery and Sadie were drowned, and his lawyers argued that the autopsies were inconclusive.

Where were they killed? All of the murders, according to the prosecution, occurred in the condominium—Zachery and Sadie were already dead when they were dropped from the bridge. Longo maintains that two were killed in the condo and two died in Lint Slough.

When were they killed? The scenarios presented by the prosecution and the defense differed by twenty-four hours, and neither side has budged. Longo wrote that he was "certain" the murders occurred on December 18, and suggested that both Dick Hoch and the harbormaster may have been pushed by the prosecution to remember the previous day. "So many questions we will never have answers to because he is incapable of the truth," MaryJane's sister Penny said at the trial. The four bodies, she said, are buried at Bethlehem Cemetery in Ann Arbor beneath a gravestone that does not bear a day of death.

FORTY-ONE

WHY?

The day after the trial ended, before I began the drive back to Montana, I visited Judge Huckleberry in his chambers. He'd heard murder cases before—he'd been a judge for more than twenty years —and I thought he might have some insight into why the Longo killings occurred.

I asked him for his opinion on the motive, and Huckleberry took in a long, gradual breath and wrinkled up his face. He said that in his career, he'd listened to the details of many horrific crimes, and all of them had a rationale, no matter how misguided or debased. But this one, he told me, was the first he'd ever experienced that seemed wholly inexplicable. The murders appeared to have happened for no reason at all.

"No behavior I've witnessed in my life fits the pattern of this case," Huckleberry said. "There's no cause and effect, no provocation, no A to B." He was fiddling with a Rubik's Cube as he spoke, twisting the parts with practiced precision—one time, during a break in the trial, he'd entertained the jury by solving the puzzle in a matter of minutes. But then he put down the cube and looked at me, clearly distressed, and held up his hands as if in surrender. He shrugged. "It's a mystery," he said.

Paulette Sanders had argued for the prosecution that Longo

had killed his wife and children because they'd become "inconvenient"—they sapped his money and his energy; they prevented him from traveling the world and achieving the glamorous, footloose existence of his dreams. "He couldn't have the life he wanted to have," Sanders said. So he killed his family in order to be free.

Stephen Scherr, Longo's psychologist, felt that Longo adored his family and wanted them to live in comfort and privilege. But when he turned them into hoboes, bouncing from place to place, evading the authorities, his pride—always a volatile and delicate element in a narcissist—was crushed. "I believe," Scherr wrote in his report, "his inability to get away from the pain and distress that he caused his wife and children led to his ridding himself of them."

I asked Longo himself, in one of the letters I wrote after his trial, to explain the motive. "It's senseless," he answered. "It should never have happened & I'm fighting every day to not continually try to figure that out. I don't think that I'll ever know."

My theory is that Longo became so entangled in his lies that he concluded murder was the only escape. He had admitted in court that he would rather steal than accept welfare. To this I would add that he'd rather kill his family than have them discover what a fraud he really was. I agree with the prosecution that the murders took place on the night Longo ate his cheese and drank his wine—that was when he realized he'd reached the limit of his deceptive abilities.

I don't think Longo ever confessed his sins to MaryJane. The all-night conversation with her that he claims to have had is, I feel, no more than wishful thinking. I believe that Longo never considered abandoning his family or allowing MaryJane and the kids to peacefully leave him. He might see them again one day, and that would be too humiliating to bear. So he had to kill them. I don't think he ever came close to committing suicide instead of murder. He believed, I feel, that if he weren't around, his family would be even worse off.

I support the defense's contention that MaryJane and Madison were killed in the condo, and Zachery and Sadie were dropped alive from the bridge. (Zachery's autopsy report indicated that the cause of death was "consistent with drowning"—a phrase the prosecution never read aloud and the defense didn't call attention to.) I do not for an instant accept that MaryJane played any role in the crimes.

If Longo had tied a sturdier knot around Zachery's leg, he might still be a free man. In an early letter, while he was maintaining his complete innocence, he said that if he were acquitted of the charges, he'd "restart on life." And if his son had remained submerged, that's what I think he would've done—restart. He would have followed through on his identity change, adopted someone else's name and Social Security number, moved to another town, taken a job, and undoubtedly found no shortage of women who were willing to settle down with him. Maybe he'd have begun a family.

I believe that Longo is a genuinely personable guy, and if I'd bumped into him at a bar, I bet we could have shot a few games of pool and had a laugh. I also believe that Longo is the most dangerous kind of man—a man who can fool even his own wife into thinking he's not dangerous at all.

Despite his death sentence, it's likely that Longo has plenty of life ahead of him. Oregon is sparing in its enforcement of capital punishment; only two prisoners have been executed since the state legalized the death penalty in 1984, and both had voluntarily abandoned their appeals. In the course of the appeals process, Longo will have the opportunity to overturn his conviction seven times. There are twenty-six people currently on "the row," as he calls it, who were sentenced to death before him.

Longo claims that he's settling into his new life as a condemned man. He exercises a lot, watches TV, and reads whatever he can get his hands on—the first book he finished after moving to the penitentiary, he said, was *The Green Mile*, Stephen King's multivolume

novel about death row. He's confined to his six-foot-by-ten-foot cell
for twenty-one hours a day, and when he is allowed out, he's sometimes handcuffed and attached to a tether, like a dog on a leash. He
works as a janitor on the row, for a salary of about $1 per day, which
he uses to pay for snacks and writing supplies. As always, he has no
shortage of admirers and pen pals. In the three months following
his death sentence, he received two marriage proposals. He corre-
sponded with his parents and brother, and said that his bond with
them was stronger than it has been in years.

"I think a lot about MJ and the kids," he wrote in a letter to me.
He's dwelled endlessly, he added, on the little things he could have
done differently, and wonders which of them may have prevented the
downward spiral that led to murder. He wants to fix what he's done
wrong, to help those he's hurt, but he knows that both of these goals
are impossible. "I feel stripped bare & don't know what to do to cover
myself up w/ a better version of me, w/o the lies & the need to feel
important," he wrote. "I guess that I feel a little lost & directionless."

He said he thinks about me, too—"pretty much every day." Even
so, we scarcely communicate. He phoned me one time from death
row, a few days after he arrived there. We spoke for fifteen minutes,
and agreed that our contact would continue only through the mail.
Over the next six months, he sent me a couple of short letters, and I
responded with brief notes of my own. Then the letters stopped,
too. Still, I'll admit that Longo is on my mind most days as well. I
can't help but wonder which version of himself he's presenting to
his fellow convicts, and how he plans to cope with the rest of his life,
and what thoughts are floating through his head.

The last time I saw Longo in private was when I visited him in
the Lincoln County Jail toward the end of his trial, on a Friday
morning when court was not in session. He knew that the trial
wasn't going well, and he was in a dour mood. Almost as a reflex, I
reverted to my lifeline role and tried to provide him with a diver-
sion. Rather than analyzing the events of the past few days in court,

I instead told him about an idea that had occurred to me while I'd been running along the Newport beach.

It had been a beautiful, cloudless day, and I'd kicked off my shoes and socks so I could feel the sand between my toes as I ran. My mind drifted into a peaceful reverie, and this vision came to me—a vision so wonderful and right-feeling that I kept running, far longer than I'd planned to, unsure if the notion would persist. But it only intensified, and the next day it was stronger still.

The vision, I told Longo, was about Jill. I'd imagined that the two of us were in Alaska—I had been there on assignment several times, though never with her—and it was near the summer solstice, when the sun is visible almost around the clock. There was snow on the high peaks, and the valleys were that brilliant, electric green that's a hallmark of the northern summer, and we were walking arm in arm along the water, and I had a ring with me, and I dipped to a knee. This is what I told Longo—I told him before I'd told anyone else—and he grinned and said it sounded superb.

On June 19, 2003, nine weeks after Longo's trial had ended, I took Jill on a surprise trip to Alaska. At midnight, on a rocky shoreline, with the sky just fading to pink, I proposed marriage. She said yes. I had fallen in love—I was getting married—in spite of Chris Longo, in spite of having spent more than a year in constant contact with a man who'd murdered his wife and children. But also, I have to admit, partially because of him, too. Working on the Chris and Mike Project had kept me at home for a long enough stretch to nurture a genuine love affair, and to become at least somewhat comfortable with the idea of settling down.

I don't know if Longo and I really grew to be friends, but to me, the bond we forged, despite the manipulation on both our parts, felt genuine. And as much as I'd like to deny it, the truth is that I saw some of myself in Longo. The flawed parts of my own character—the runaway egotism, the capacity to deceive—were mirrored and magnified in him. All the time I spent with Longo forced me to take a lengthy and uncomfortable look at what I'd done and who I had become.

My year with Longo made me see how a person's life could spiral completely out of control; how one could get lost in a haze of dishonesty; and how these things could have dire consequences. I believe that if I had not met Longo—if I'd tried another way to revive my life after the *Times* disaster—I may have learned similar lessons, but not so quickly and clearly and profoundly.

From the first week I was fired, I knew that I owed an apology to my editors, to the fact-checkers, to the photo department, to my colleagues, and to everyone who had read the West Africa article. But I didn't feel that an apology from me would be accepted as genuine. Now, I hope, it will. And so the last thing I want to say about my *Times* article is this: I'm sorry.

In one of the letters Longo wrote me from death row, he asked if I'd followed through on my Alaska idea. I responded that I had. He wrote back, and said that the news made him feel "warm & fuzzy inside, happy; a feeling that I haven't felt in a while." He assumed it was only a matter of time before babies were on the way, and he gently teased me, visualizing me walking around town "wearing one of those backpack kid carriers." He also congratulated me on finally becoming an adult, at age thirty-four.

Soon after my engagement, I began assembling all the pieces I'd gathered for the Chris and Mike Project, organizing them into a book-length manuscript. To put myself in the proper frame of mind, I reread all of Longo's letters one more time, in chronological order.

Something in the opening section of his very first letter—the one detailing his time in Mexico—made me laugh. Longo was aware, right from the start, it seemed, that I would never figure out what really happened in Lincoln County in the early hours of December 17 or 18 of 2001. By the time his trial ended, the amount of information on Longo that I'd examined, including legal materials, transcribed conversations, and media clippings, had exceeded five thousand pages. But in his first letter, on page 13, he noted that "no matter how many thousands of pages" I eventually compiled, "they would never provide an accurate enough description, or explanation, of the entire story."

Twenty-nine letters later, after Longo was imprisoned on death row, he counseled me on another issue. In the whole of my book, he stated, "any inaccuracies, regardless of how innocuous" would drive him "nuts." This was written on the one-thousand-one-hundred-forty-third page he'd mailed me. Those pages included three different versions of the murders—one in which Longo was responsible for none of the killings, another in which he was guilty of two, and a third in which he admitted to all four. Yet he insisted that the story I tell must be a model of precision. He won't be pleased, he said, unless everything in this book is absolutely, unassailably true.

EPILOGUE

LONGO AND I did not speak for eleven months. There was one brief call just after he'd been sent to death row, then silence while I worked on this book. On the very day that I wrote the final sentence of the first draft—March 27, 2004—my home telephone rang. Sure enough, there it was on my caller ID: INMATE PHONE. Instinctively, I answered. Longo's familiar, even-keeled voice was on the other end of the line. He was calling, he told me, to announce some big news.

"I'm engaged to be married," he said.

Over the course of our communication, Longo and I had spoken on the telephone for a total of fifty-one hours, and in all that time I'd never once been struck mute. Now I had no idea how to respond.

After his incarceration on death row, Longo explained, he'd received dozens of letters, and had replied to a handful. In one instance, the letter writing burgeoned into a full-blown epistolary romance. Soon enough, Longo said, he was calling and writing to this woman every day and seeing her in twice-weekly visits of two hours each, the maximum he was allowed.

Finally, Longo informed me, he had proposed marriage. The woman said yes—then went out to buy her own diamond ring. As I listened to Longo wax poetic about his newfound love, I did the math in my head: It had been twenty-seven months since he'd stuffed MaryJane and Madison into suitcases and dropped them in Yaquina Bay, then tossed Zachery and Sadie off the Lint Slough bridge.

In 2005, a year after Longo told me of his engagement, I returned to Oregon for the first time since the trial. *Vanity Fair* magazine was printing an excerpt of this book, and I was attempting to persuade the authorities at the Oregon State Penitentiary to allow Longo to be photographed for the article.

My request was denied, but I was permitted to spend two hours with Longo. In order to see an inmate on Oregon's death row, a visitor must first pass through what's known as the "tiger run," a chain-link contraption that does indeed resemble the elongated set of cages used to transport circus tigers from backstage to center ring. I entered the run, escorted by a guard who unlocked and re-locked the various doors, passed through the penitentiary's two-story-high cement walls, and ended up in the visiting room.

Longo was waiting for me. As in the Lincoln County Jail, we spoke over telephone receivers through bulletproof glass. He sat on a metal stool bolted to the floor and I sat in a cushioned office-style chair on rollers. I hadn't seen him in two years. He looked as youthful as ever, his hair precisely arranged, with a little triangular patch of stubble beneath his lower lip.

He'd bulked up considerably, a result, he said, of his daily exercise regimen. He was wearing a light-blue collared shirt, washed-out blue jeans, and white New Balance sneakers. I commented that his outfit didn't seem particularly inmate-like, and he told me, practically bragging, that on death row you didn't have to wear the bright-orange uniform of the general inmate population.

I asked about his fiancée. Of all the letters that were written to him, why had he answered hers? Longo told me that she'd included a photo with her letter, and that he hadn't found her attractive. He thought, therefore, it would be safe to respond. As it turned out, he added, the photo happened to be a terrible one of her. He was also drawn to her bluntness; she told him in her very first letter that she didn't ever want him to meet her two children—and, he said, he never has.

When I mentioned the marriage proposal, Longo demonstrated how he'd asked for her hand. He stood up and placed his right knee on the stool; if he'd have kneeled on the floor, he explained, he would have disappeared beneath the window. Their wedding, Longo said, had not yet taken place—the prison authorities were in no rush to organize such a ceremony.

Longo said during my visit that he seldom dwelled on the murders, and if he felt his mind drifting in that direction he consciously pinched off the thought. Only on the anniversary of the killings and on the birthdays of his children was he really pained. As for his appeals, he wrote this in a letter he sent me after my visit: "I've decided to pursue my appeals full-force . . . to ensure that I will provide happiness to my wife for as long as I possibly can, and of course to enjoy a life with her as well." The depression he felt when he was first incarcerated, he told me, had fully dissipated.

"Life has been as close to wonderful as a person could ask for, and I have no reason to not savor a smile," he wrote. He went so far as to imply that it was worth murdering his family in order to have met this new woman. "I just feel as though I am extremely fortunate to have fallen into the right spot, almost like it's exactly what was supposed to happen." He and his wife-to-be, he later added, "really are the perfect match." This new match, he suggested, was far superior to the one he'd had with MaryJane.

"I have another family," he wrote me, one that "means everything that my own children meant to me, and one that I would do anything to protect."

Our death-row visit concluded with an odd request. Longo said that he not only wanted me to meet his fiancée, but that he'd also pay for our lunch. I didn't even ask how he'd manage to pick up the bill. I did, however, agree to see the woman, and I phoned her after the visit. We decided to meet at the Coffee House Café in downtown Salem, a few miles from the penitentiary. I arrived first, sat at a window table, and ordered a pot of tea.

Soon enough, she walked in. She was a pretty, petite woman, with long blond hair, blue eyes accented with a touch of glittery makeup, and a tiny stud in one side of her nose. Her perfume was strong and sweet.

I asked what had motivated her to write the initial letter to Longo. She claimed that it was the first time she'd ever written an inmate. She was drawn to him, she said, by some inexplicable force. She'd hadn't really followed his case, but she did know he wasn't originally from Oregon, and said she felt sorry for him, being out here all alone, with no family around. She told me that she lived forty miles from the prison, but didn't mind making the drive twice a week. She said she especially loved his letters; he'd written her an average of four pages a day for more than a year. "He's an awesome writer," she told me. She was amazed by his vocabulary; one of her nicknames for him was "Mr. Dictionary."

I could not understand why this woman, attractive and well-spoken, would become so involved with a man on death row. It seems possible, however, that in many ways her relationship with Longo felt like the ideal scenario. I don't know much about her past relationships—we met only this once and I didn't pry too deeply— though she did reveal that the father of her two children was also in prison (not the same one as Longo), serving a six-and-a-half-year sentence on break-in and attempted rape convictions. In her relationship with Longo, she's completely in control; she determines whether to answer the phone, or write a letter, or arrive for a visit. Longo, meanwhile, relishes her attention and sends her constant love letters. And there's never any sexual pressure: The two can't even touch. The wedding between Longo and his fiancée, if it ever happened, would be a ten-minute ceremony in the visiting room. No conjugal visits allowed.

Longo's fiancée seemed entirely serious about her intention to wed Longo. She proudly showed me her engagement ring—it was modest and tasteful. She was also, I understood, very protective of the relationship; Longo had mentioned during our visit that she grew jealous whenever he received letters from other women. Evidently, though, this didn't deter Longo from carrying on a fairly intense correspondence with another woman. Three months after my meeting with Longo's fiancée, this second woman showed up at a reading I gave in Portland, Oregon. She brought with her a stack of two dozen letters Longo had recently written her. I skimmed them for a few minutes. They were flirty and friendly, with Longo's typical attentiveness; in one, he sketched designs for the logo of a coffee shop she was thinking of opening. When I said that Longo was engaged to be married, she was shocked. In those two dozen letters, Longo had never once mentioned his fiancée—he'd actually hinted that he was lonely and rarely had visitors.

The last thing I asked Longo's fiancée, before we parted, was what Longo had told her about the murders. It seemed, from her story, that he was once again blaming MaryJane for killing Zachery and Sadie—though Longo's fiancée, I noted, never spoke the name MaryJane aloud. She simply referred to Longo's first wife as "she" or "her."

In a letter Longo wrote me after our death-row visit, he insisted that, despite all he'd told me, there were still "blanks to be filled in" about the murders. In this same letter, he disclosed what he planned to give his fiancée as a wedding present: "a book-length reveal of the utter truth of everything, from beginning to end . . . with all the blanks filled in."

And so, it seemed, there would be at least a fourth explanation of the crimes. This one, Longo insisted, would be the *real* true story.

* * *

As for this *True Story,* the book itself, Longo was eventually permitted to read it. I'd mailed him a copy as soon as it was printed, but it was not delivered to his cell. Books written about inmates at the Oregon State Penitentiary are not allowed into the prison. However, because this book contains confessions that were made after his trial—confessions that could affect his appeals—Longo and his appellate attorney petitioned the prison to allow him to see it. He was successful, and was able to keep it in his cell for a couple of weeks.

Soon after Longo read it, in May of 2005, he phoned me, then wrote a follow-up letter. He said he'd read the book three times. *True Story* did, Longo wrote, "relay honestly what was said between us," but what bothered him, he said, was "the tone." I asked for more specifics, and he said he was "shocked" I'd written that he was using our correspondence as a dress-rehearsal for his testimony. I had portrayed our whole relationship as "a long con, more or less" and, he said, "that's what killed me. That's the only thing that I even really have a major contention with."

I wasn't surprised. I knew Longo hated to be called a con man. He'd always maintained that all the crimes he committed before the murders were strictly to help his family, and whenever a newspaper labeled him a con artist, he'd complain to me over the phone or in a letter. And now I'd done the same thing to him.

In one of the letters that Longo's other woman showed me, he mentioned my book. "The author made a lot of his own conclusions that were pretty off," Longo wrote. "But I can see how he came up with them. And he turned out to not be the type of friend that most would want to divulge so much to. But he is a journalist after all. Duh!"

To me, he wrote this: "Overall I regret contacting you." He regretted it, he said, because my book had caused additional pain for his family and MaryJane's family by forcing them to relive the crimes. I have not heard from either family. Still, Longo concluded his letter by saying that he didn't "harbor any resentment or ill-will at all." Rather, he added, our relationship was by no means over. "I would like it very much," Longo wrote, "if we continued to be in association with each other."

We did not continue, at least not for a while. Three years passed with little more than an exchange of holiday cards. Longo did not, after all, get married; his fiancée apparently came to her senses. The breakup occurred when she faced a series of difficult circumstances—health issues, financial concerns, the death of her father—and Longo could do nothing but listen over a monitored prison phone or behind a pane of thick glass. "A landside of life's dramas with only a figurative shoulder to lean on," is how Longo put it. This wasn't enough to keep a relationship going. And so it ended, though the couple remained close friends.

In the meantime, my own life blossomed. Jill and I, who were married in Montana in 2004, started a family. Our first child, a girl, was born in 2005. Nineteen months later, Jill gave birth to a son. I was a family man and, once more, a working journalist— several publications were willing to hire me, though not the *New York Times*. Between the babies and the work, I seldom thought about Longo, but there were always moments, I admit, when I pondered what he was up to.

Jill was eight months pregnant with our third child when, on the evening of February 21, 2009, the telephone rang and there it was again: INMATE PHONE. I remember thinking, as I picked up the receiver, that there's no way Longo would be able to deliver news as disturbing and unexpected as an impending marriage. But I was wrong.

He asked me if I wanted to watch him die.

The reason he wanted to die, he said, was that after half a decade spent sealed inside a white concrete box, with only other murderers as neighbors, and with no chance of ever again seeing the outside world, he'd had enough. The joy he'd expressed in the wake of his engagement had vanished. Now there was no fiancée and, it seemed, no hope for the future. He was sick of prison and sick of himself, and he thought there might be a way to make his death meaningful. So he was dropping his appeals, he told me, and would likely be executed, by lethal injection, in less than a year.

"I'm not going to make it to my thirty-sixth birthday," he said. A Will Smith movie, he explained, changed him. He saw it on the seven-inch flat-screen TV inside his cell, a picture called *Seven Pounds* about a guy who's so distraught after killing his fiancée and six others in a car accident that he decides to commit suicide and donate his organs to people in need.

The movie, Longo said, felt like a punch in the gut. It made him weep. For more than a year, he said, he'd sat in his cell wondering how he could do anything worthwhile, anything at all to help even one person, rather than just rot away on death row. The movie gave him the answer. He would carve himself up. He'd give away his heart and lungs and liver and corneas and bone marrow and whatever else could be salvaged. He could singlehandedly save the lives of as many as eight other people. His "finale," he called it. Let others live; let him die. That's what he wanted.

He'd reached this conclusion, he told me, after conducting a strange, self-administered psychological test. For the first time since he'd been incarcerated, he hung up a photo of his children in his cell. It was a studio shot, one I'd seen at his trial, the three kids gazing smiley and wide-eyed into the camera, heartbreakingly cute.

"Every time I turned around or rolled over, there they were staring at me," Longo wrote in a letter he mailed me soon after we spoke on the phone. But his reaction to the photo distressed him. "I'm not really feeling what everyone else feel's," he wrote, tossing in, as he often did, an extra apostrophe. "What should be most difficult to stomach is what I've done, yet somehow that part is still palatable."

Lately, he added, when he looked in the mirror he was "beginning to see a monster." The best solution, he'd concluded, was to give away his organs and "end on a good note."

But first he needed a favor. That's why he was calling. He asked if I'd be willing to help him formulate a plan to donate his body parts. His former fiancée, he added, was also working on the project, but he needed a journalist's skills to solve a few problems. I said I needed some time to wrap my mind around the idea. But already I was thinking that a world without Longo in it wouldn't be a tremendous loss. It might give MaryJane's family some peace, and maybe his own too. So I asked him what he wanted me to do.

"I'll write you a letter," he said.

And just like that, it started again. Longo never does anything halfway. He didn't merely send one letter. He began mailing me a letter every week—lengthy, intricate, sometimes disturbing letters. Many requested my help; the Oregon State Penitentiary forbid prisoners from donating organs, and Longo hoped to change this rule before being put to death. This involved petitioning state legislators and prison officials and organ-transplant experts, among others, and soon I was lost in a bureaucratic labyrinth. Every person I got in touch with seemed to insist that I contact two more.

Longo's letters also included detailed glimpses of his strange life on death row. His existence, like those of Oregon's other condemned men—there are currently thirty-four people on the row—was largely confined to a windowless cell. (One woman and a man who requires dialysis are incarcerated elsewhere.) No one had a cellmate on death row, and everyone was permitted his own television. When allowed to leave his cell, he played basketball twice a week with three other death-row inmates, and lifted weights regularly. His biggest regret was a lack of Internet access.

Longo kept his cell spotless and meticulously ordered; every

spice bottle, every jar of lotion and shampoo, every book had a particular spot. He once listed the novels on the shelf above his bed:
Anna Karenina, *The Inferno*, *The Metamorphosis*, *Candide*. Other cells,
he wrote, were littered with "remnants of the previous meal, salt &
pepper flakes everywhere, a half dozen roll's of partially used toilet
paper." Still, there were general rules of etiquette. It was polite, for
example, to mask particularly audible flatulence by simultaneously
flushing your toilet. An odor repellant was also essential. Inmates
tore out scent strips from magazines or spread deodorant on their
bars.

Some inmates, whom Longo called "slugs" in his death-row tax-
onomy, essentially did nothing all day. They ate, watched TV, slept.
The "kids," on the other hand, were generally "argumentative, dis-
respectful of everyone, short tempered, dirty." The "matures"—the
group Longo included himself in—obeyed the rules and pursued
prison-appropriate hobbies like playing chess or learning the guitar
or creating elaborate pen-and-ink drawings. Longo said he spent so
many hours in his cell doing sit-ups, push-ups, and toilet-seat step-
ups that his resting heart rate was as low as forty-six beats per minute.

Everyone, though, seemed to agree on the importance of porn.
On Oregon's death row there are never contact visits, so for the rest
of your life you can't so much as touch a visitor's hand. Longo's
porn stash, he wrote, was relatively tame: "Where I might be content with the Playboy-esque spreads, somebody else needs something involving fists." The inmates sold it to one another; the going
rate was between fifty cents and a dollar per page.

The second-most prevalent obsession was food. Longo said
he actually had two photo collections: nude women and gourmet
cuisine. To make the institutional meals more palatable, the men
sometimes held death-row dinner parties. Several inmates would
pass their trays down the row to one cell—frequently, to Longo's.
He'd combine all the food together, add items ordered from the
prison commissary (and delivered to his cell once a week) like hot

sauce, peppers, and shredded cheese, then rebuild the plates "Cadillac style," as it was called, and send the trays back.

Real death-row parties required a batch of pruno, the prison hooch. Longo sent me his recipe. First, take the grapefruit that comes with Saturday breakfast, peel, crush, and place in an empty milk carton. Enzymes remaining inside the carton, combined with natural yeast from the air, trigger fermentation within a few days. Dump into a garbage bag and add a pound of sugar and a pound of commissary-bought prunes per person. Let sit for a week, occasionally—when no guards are around—burping out excess gas. Strain pulp through a sock. The alcohol content, Longo said, was as much as twelve percent, and the taste was about what you'd expect from something filtered through a sock. Still, if you could stomach a couple of twenty-ounce cups, pruno provided "a good escape."

Generally, according to Longo, everyone on death row got along, at least on the surface. "There's a weird pseudo-cordiality thing here," he wrote. "But you know that given the opportunity to get away with it a few here would have no compunction about stabbing someone where they sleep." In truth, he wrote, nobody truly trusted anyone. Several men were transferred to death row from the general prison population after murdering other inmates. A few were serial killers. The death-row barber, Dayton Rogers, was convicted of killing six women (though he may have murdered more), and liked to cut off the feet of his victims with a hacksaw while they were still alive. "I'm surrounded," Longo wrote, "by so much degeneracy and perversion." I once asked if he thought anyone on Oregon's death row was innocent. "No," he replied.

In some ways, Longo said, you got to know the other people on death row extraordinarily well—their favorite fishing hole, the name of their childhood pets—but it could be challenging to separate their real lives from the fantasy version. One inmate told Longo, in intricate detail, how he'd scuba dived in Lake Michigan to see the shipwreck *Griffin*. But Longo had just watched the same show about the *Griffin*, on the Science Channel, from which the inmate had cribbed every fact. The truth, Longo wrote, was that on death row "nobody is what they seem."

This included Longo himself. Money is an important asset in prison—not only to purchase snacks and supply your cell with the best possible television and handheld video game console but also to garner status and respect. Longo's facade in prison was the same as it was in the outside world: a successful businessman. On death row, people thought he was a stock-market whiz. And on the surface he seemed to be. He subscribed to the *Wall Street Journal* and *Barron's* (using frequent flier miles left over from his former life) and often kept his TV tuned all day to CNBC. He supposedly called his broker with picks and earned big profits. It was actually an elaborate ruse. "All of that pretend stock market playing is believed to be real," Longo wrote. "I've never told anyone that it's not. And I use the phone for sufficient amount's of time to all for that thought to seem legit."

Longo was, indeed, making money on death row. But not on the market. He was really providing titillating letters and phone sex to a couple of gay men. This is a rather common fetish, it turns out; whenever a new inmate arrives on death row, according to Longo, he'll receive a pile of letters from men who've followed the case through the media, seeking a prison lover, perhaps turned on by the thought of an amorous murderer. Such men—sometimes referred to as "ATMs"—will repay the letters and calls with generous deposits in a prison account.

Longo had two ATMs. One was an accountant in San Francisco: "He saw me as the chick in the relationship and constantly tried to force the wife title on me." The other was a schoolteacher (and grandfather of three) in New York City: "His fantasy was to have me crucified, Christ style, so that he could pull me down and tend to my wounds." He had these two "on the hook" for more than five years; he claimed they'd given him thousands of dollars.

Maintaining the stock-market lie, Longo wrote, was getting "exhausting." But he couldn't be honest, he explained, because of "extreme embarrassment." It was a similar combination, I believe— too tired to keep lying, too ashamed to reveal the truth—that led him to murder his family. Only this time, he wanted to kill himself.

"I'm in a place," he wrote, "where you're surrounded by similarly messed up people with only one mission. To do time. Something I'm more than worn out on already." There was really no point in continuing, he said. He was ready to carve himself up. "After looking at my neighbors who have been here for a dozen years or 30, it's all an exercise in futility. So you can play the guitar well enough to be in a band & entertain millions. Or you've studied magic as a hobby & now you can dazzle the group . . . Five, ten, fifty years later you're still living in a box amongst a bunch of other boxes within a bigger box."

As the months passed from the time Longo told me he was planning to die, he explored, over the course of a dozen dense letters, the peculiar environment of death row and, more intimately, the complicated landscape between his ears. The only time he'd been thoroughly examined by a psychologist, shortly before his trial, the report labeled him a borderline psychopath.

Longo did not disagree with the assessment. But he did debate precisely which side of the line he was on. After hanging up the photo of his children and hardly reacting, he called himself "dead inside" and "more cold-hearted than I could ever imagine a human could be." Other times, he was convinced he was just as capable of emotion as anyone else: "If I am a monster how could I have ever loved them as I felt I did?" he wrote about his family. "It didn't equate." Also, he noted, as further refutation of his psychopathy, "I got choked up during E.T. & Titanic."

The longer he thought about all this the more confused he became—he said he felt "nearly constant guilt over not feeling guilty enough for what I did"—and his attempts to understand his behavior seemed "like trying to unscramble really rotten eggs." Eventually, he gave up. His letters turned darker and more fatalistic. "It does feel very much," he wrote, "like the day's are ticking by time-bomb fashion." Longo agreed to put me on the witness list for his execution, and said he'd be willing to meet with me on the week leading up to his death. I found myself wondering what he'd order for his last meal, or say in his final statement, while strapped to the gurney.

Then my wife stepped in. Jill has always been sickened by Longo's crimes, but now her distaste was even more acute. In March of 2009, a month after Longo had informed me of his organ-donation plan, she gave birth to our third child, a girl. The fact that I, a man with a wife and three young kids (two girls and a boy), was speaking with a man who killed his wife and three young kids (two girls and a boy) greatly unsettled her. She was particularly upset whenever Longo bantered about my fatherly duties—"I'm sure," he wrote, "there's a diaper that needs changing, or a car seat to clean out. I know."

And yet my apparent enthusiasm for, and ancillary participation in, his execution also troubled her. She felt, in essence, that I was helping kill a person with whom I'd had a profoundly intricate relationship, and that this would weigh on my conscience. She said what I first needed to do, before I agreed to witness Longo's death, was try my hardest to talk him out of it.

I realized she was right. So I wrote Longo a letter. I told him that I believed he really wanted to "enhance someone else's life," as he'd written, by sacrificing his own, a real-life version of the Will Smith movie. But there were a couple of major problems. First, the prison hadn't changed its rules: Organ donation was not allowed. And even if it was permitted, I'd learned that the Oregon execution procedure—sodium pentothal, then pancuronium bromide, then potassium chloride poured into the veins—rendered all organs useless. Some skin tissue might be saved. Maybe the heart valves. Then his body could be donated to a medical school.

It didn't seem to fulfill his goal, and I told him so in my letter. The problem, I wrote, was that the state-administered death cocktail produced heart failure. If you were able to change the procedure (the law isn't specific about the precise drugs used) so that it induced brain death instead, the organs could be transplanted. Or a large dose of just a single barbiturate, sodium thiopental, which is used in Ohio and Washington state executions, might do the trick. Then work on modifying the prison's organ-donation rules. And finally, I continued, you could sign up other inmates, and perhaps take the idea to a national level, and thereby potentially save the lives of dozens or even hundreds of people who would've died on organ waiting lists. "Sounds like 10 years work to me, minimum," I wrote, though I also noted that he could quit, any time, and resume his current plan. "What's your reaction?"

Longo was astounded. When he read my letter, he told me, something inside of him clicked. He felt an enthusiasm he hadn't experienced since his prison engagement. He felt inspired. It's "giving me goose bumps," he wrote.

And so, in his single-minded way, Longo promptly dedicated himself to making the idea a reality. He came up with a name: GAVE. It stood, at first, for Gifts of Anatomical Value from the Executed. Later he changed it to Gifts of Anatomical Value from Everybody. With the help of his brother, Dustin, and his former fiancée, he set up a website, GaveLife.org, and a Facebook page (G.A.V.E.). "I am a death row inmate who wants to save lives," Longo wrote on his site. "Not to set right my wrongs—as this is unfortunately impossible—but to make a positive out of an otherwise horrible situation."

* * *

Longo embarked on a flurry of letter writing and research, poring through law journals, gathering statistical data. He contacted law schools, hospitals, legislators, attorneys, organ-donation groups, transplant officials, and anesthesiologists. "I'm up at 5:05 a.m. every day," he wrote, "and rarely knock out before midnight and the entire parts in the middle seem to vanish way too quickly." He constructed and mailed me an intricate eight-page outline of things he had to accomplish to make GAVE a success. There were more than three-hundred separate bullet points and eighty-five questions that needed answers, covering everything from inmate transportation to hospital security to the cost of follow-up care.

The list overwhelmed me, and must have overwhelmed him too. He'd seemed driven and focused and confident, though Longo almost always seemed that way on the surface. He said he was no longer dropping his appeals. I assumed he was satisfied, and that the GAVE project, in a way, was able to transport him beyond the prison walls.

But I had misread him. In prison, Longo was prescribed a drug called amitriptyline, an antidepressant that also helped alleviate Longo's occasional migraines. For three months, rather than swallow the pill, he crushed it up and hid the powder in a jar of protein mix. When he had three ounces worth—more than twice what he figured he'd need (he had carefully studied the drug's effects in overdose quantities)—he was ready. It was just past New Year's Day, 2010. He spoke with his ex-fiancée on the telephone for two hours. She apparently knew what he was going to do, and Longo wanted to be certain that all of his medical directives and power-of-attorney rights were in proper order. Then they made their peace with each other. "We got to a good place," Longo wrote. When he hung up the phone, he said, they were both crying.

He poured all the medicine into a thick drink of cocoa, coffee, and sugar. "My thought," Longo wrote, "was that I would never be able to get the prison to let me donate." In terms of legalizing organ donations from executed inmates, he felt it would require all his efforts "to move 10 feet of a miles-long journey." Better to just take a short cut. While he mixed his concoction, he wrote, he was "feeling certain that everything was in order to at least ensure some organs survived long enough to be rescued, fully expecting to not wake up from the medical coma." He drank it down.

Longo woke up in Salem Hospital, very much alive, with all his organs still inside him. The prison had ignored his medical directives. "They did everything they could to revive me," he wrote. "Gotta keep me alive to kill me the right way."

But the suicide attempt, in an odd way, gave him renewed energy. Within weeks of trying to kill himself, he was once again working on his GAVE project, with greater vigor than ever. He soon expanded the mission. His hope is that every inmate, not just those on death row, will be allowed to make living anatomical donations—a kidney, liver lobes, bone marrow—and then those who are executed or who die of other causes will be able to give away all the major organs. More than a hundred thousand Americans are on organ-transplant waiting lists. About thirteen people die every day waiting for an organ. Over two million people are incarcerated in the United States, an enormous pool of potential donors.

Recently, some startling progress has been made on the issue. On March 28, 2013, Utah governor Gary Herbert signed into law a bill allowing inmates to become organ donors if they died in prison. It was the first law of its kind in the nation. Seven other states, though not yet Oregon, are currently considering similar bills. And in November of 2013, Ohio governor John Kasich halted the execution of child-killer Ronald Phillips so the state could study the feasibility of executing Phillips in a way that allowed his organs to be donated. Ohio, so far, has been unable to figure it out, but the organ-donation idea is clearly on officials' minds.

"I'm not happy. I just don't despair," Longo wrote me recently. "I operate on ambition, as I think I always have." He says he will keep pushing forward on organ donation. He says he has no plans to try another suicide attempt, but won't definitively rule it out either. Living on death row, he wrote, is "a lonely, haunting existence, as it probably should be."

In January of 2014, Longo turned forty years old, three weeks after I turned forty-five. I've now been in contact with Longo for a quarter of my life. An "unorthodox journey," he called our relationship. His oldest child, Zachery—murdered before his fifth birthday—would nearly be seventeen, likely driving a car, finishing high school, on his way to adulthood. Longo wrote that he dreams about his kids often, "always w/ them in distress, always w/ me to blame." But he still feels that "we were a great family." He also finally admits that MaryJane had nothing at all to do with any of the murders. He, alone, killed his entire family.

Longo and I remain in touch. He usually phones me on the first Sunday of every month, chiefly so I can navigate the Internet and help with his organ-donation quest. This book is being made into a movie, and Longo hopes to use the publicity that accompanies the film's release to bring more attention to the issue of prisoner organ donation.

"I am seeking nothing but the right to determine what happens to my body once the state has carried out its sentence," Longo wrote. He composed that line about a year after his suicide attempt, in an impassioned essay describing his cause. He mailed the essay to me for editing, then submitted it for publication to several newspapers.

One newspaper accepted Longo's essay. It was, in fact, a rather prestigious paper. On Sunday, March 6, 2011, beneath the headline

"Giving Life After Death Row," Longo's article, with his byline on it, was prominently featured on the back page of the Week in Review section, and he officially became Christian Longo of the *New York Times*.

—Michael Finkel
Bozeman, Montana
July 16, 2014

ACKNOWLEDGMENTS

OVER THE THREE years that I worked on this project, my agent, Stuart Krichevsky, was an inexhaustible source of sage advice and absolute support. To him I give my thanks, knowing full well that no measure of thank-yous could suffice.

David Hirshey and Mark Bryant, my editors at HarperCollins, were the ideal tag team, by turns patient and demanding, gentle and not so, and at all times devoted and visionary. I am deeply grateful for the seemingly limitless energy they expended on this book.

Paul Prince, who has served as a mentor throughout my career, offered astute and judicious guidance throughout the manuscript's construction. Nick Trautwein's extraordinary attention to the vagaries of pacing and style have saved this project from many a false note. All the remaining ones are in spots where I've likely ignored his advice.

Kevin McDonnell, who accepted the daunting task of checking the facts, performed beyond the call of duty. Rachel Elson was an ideal reader and editor—ruthless and consistently right, dammit. Adam Cohen, another ace reader, was a bloodhound for foul-smelling prose. Both made the book better.

Matt Sabo and Bryan Denson of the *Oregonian* generously shared their research with me. Karen Franklin, Joe Dixon, and Elisabeth Young-Bruehl provided insightful analyses of Longo's letters. Steven Krasik and Ken Hadley helped me grasp the dynamics of the trial. Gloria Thiede earned my nomination for patron saint of tape transcription. Dorothy Dacar offered cranial cleaning. Richard Robitaille, Bill Bishop, Andrew Kramer, Alberta Bryant, Charles Hinkle, and Lisa Harmening selflessly assisted with my reporting in Oregon.

Janet Markman, Shana Cohen, Miles Doyle, Lawrence Weschler, Zeynep Gürsel, Alison Samuel, Mark Miller, Mark Jannot, John Byorth, Larry Smith, Bob Hamilton, Doug Schnitzspahn, Steve Byers, Abby Ellin, Rachel Lehmann-Haupt, Randall Lane, Brett Cline, Mandi Hardy, Amos Blinder, Jeanne Harper, Chris Anderson, Alan Schwarz, Arthur Goldfrank, H.J. Schmidt, Kent Davis, Katie Goodman, Soren Kisiel, and Valerie Hemingway provided creative inspiration, thematic first aid, and shoulders to lean on.

A warm embrace to Eileen Finkel, Paul Finkel, Diana Finkel, and Ben Woodbeck, who put up with (and maybe even understood) my shenanigans. Thank you for your encouragement and your love. Harris and Theresa Barker have also been steadfast supporters, despite having a Yankee in their midst. Thanks also to Grandpa Manny, who taught me how to tell a story, and is in my thoughts every day.

Most of all, my wholehearted love goes to Jill Barker—now Jill Finkel—for her devotion, her spiritedness, and her unflagging sense of humor. And, of course, for making an honest man out of me.